International Conference on Experimental and Computational Mechanics in Engineering (ICECME 2019)

AA007694

IOP Conference Series: Materials Science and Engineering
Volume 931

Banda Aceh, Indonesia
18 - 19 September 2019

Editors:

Iskandar
Akhyar
Syifaul Huzni
Arhami
Husni
Muhammad Dirhamsyah
M. Nizar Machmud
Muhammad Rizal
Nurdin Ali
Rudi Kurniawan
Sarwo Edhy
Sulaiman Thalib
Syarizal Fonna
Zahrul Fuadi
Hamdani
Muhammad Ilham Maulana

ISBN: 978-1-7138-3193-8
ISSN: 1757-8981

Printed from e-media with permission by:

Curran Associates, Inc.
57 Morehouse Lane
Red Hook, NY 12571

CURRAN ASSOCIATES INC.
proceedings
.com

Some format issues inherent in the e-media version may also appear in this print version.

Institute of Physics
Dirac House, Temple Back
Bristol BS1 6BE UK

Phone: 44 1 17 929 7481
Fax: 44 1 17 920 0979

techtracking@iop.org

Additional copies of this publication are available from:

Curran Associates, Inc.
57 Morehouse Lane
Red Hook, NY 12571 USA
Phone: 845-758-0400
Fax: 845-758-2633
Email: curran@proceedings.com
Web: www.proceedings.com

TABLE OF CONTENTS

MECHANICAL AND INDUSTRIAL ENGINEERING

IOP Conf. Series: Materials Science and Engineering **931** (2020) 011001 doi:10.1088/1757-899X/931/1/011001

PROCEEDINGS

International Conference on Experimental and Computational Mechanics in Engineering

September 18-19, 2019

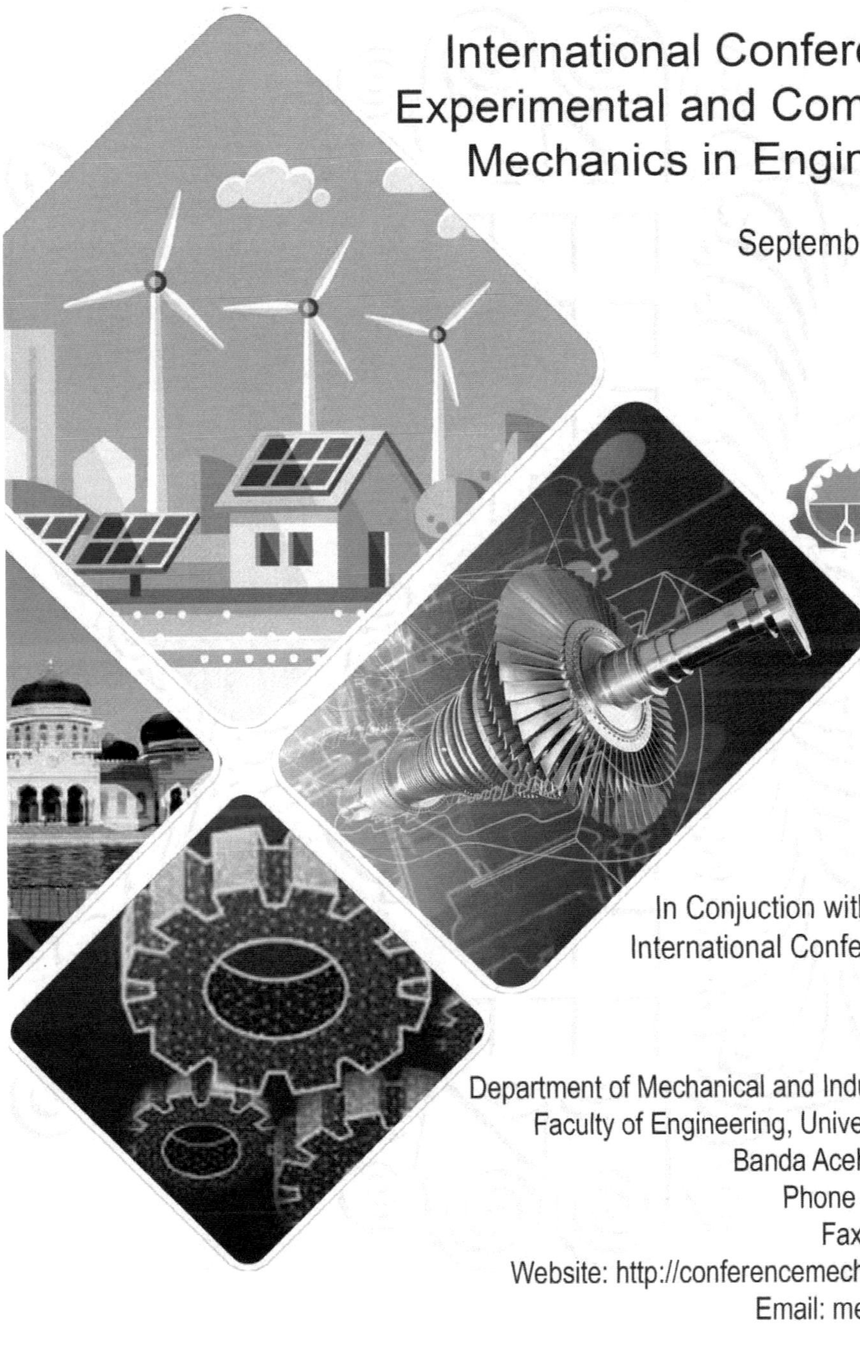

ICECME 2019

In Conjuction with The 9th Annual International Conference (AIC)2019

Department of Mechanical and Industrial Engineering
Faculty of Engineering, Universitas Syiah Kuala
Banda Aceh, Indonesia 23111
Phone +62 85361797661
Fax +62 651 7552222
Website: http://conferencemechanic.unsyiah.ac.id
Email: mech@unsyiah.ac.id

IOP Conf. Series: Materials Science and Engineering **931** (2020) 011001 doi:10.1088/1757-899X/931/1/011001

PROCEEDINGS

INTERNATIONAL CONFERENCE ON EXPERIMENTAL AND COMPUTATIONAL MECHANICS IN ENGINEERING (ICECME)

DEPARTMENT OF MECHANICAL AND INDUSTRIAL ENGINEERING
FACULTY OF ENGINEERING, UNIVERSITAS SYIAH KUALA
SEPTEMBER 18 – 19, 2019

IOP Conf. Series: Materials Science and Engineering **931** (2020) 011001 doi:10.1088/1757-899X/931/1/011001

Conference Information

Conference Name	International Conference on Experimental and Computational Mechanics in Engineering (ICECME) 2019
Conference Logo	ICECME 2019
Host and Organizer	Dept. of Mechanical and Industrial Engineering, Faculty of Engineering, Universitas Syiah Kuala
Venue	Academic Activity Center (AAC) Prof. Dayan Dawood, Universitas Syiah Kuala, Banda Aceh, Indonesia, and Kyriad Muraya Hotel, Banda Aceh, Indonesia
Official Language	English
Secretariat	Jurusan Teknik Mesin dan Industri, Fakultas Teknik, Universitas Syiah Kuala, Banda Aceh, Indonesia Phone +62 85361797661 Fax +62 651 7552222 Website http://conferencemechanic.unsyiah.ac.id/ Email mech@unsyiah.ac.id

IOP Conf. Series: Materials Science and Engineering **931** (2020) 011001 doi:10.1088/1757-899X/931/1/011001

Honorary Advisory Board

Rector of Universitas Syiah Kuala	Prof. Dr. Samsul Rizal
Vice Rector I	Prof. Dr. Marwan
Vice Rector II	Dr. Agussabti
Vice Rector III	Dr. Alfiansyah Yulianur
Vice Rector IV	Dr. Hizir
Head of Institute for Research and Community Services	Dr. Taufik Fuadi Abidin

Organizing Committee

Chairman	: Dr. Ir. Mohd. Iqbal, M.T
Vice Chairman	: Dr.-Ing. Teuku Edisah Putra, S.T, M.Sc., Ph.D
Secretary	: Ratna Sary, S.T., M.T
Finance	: Dr. Ir. Razali, M.Si, M.T

Committee Members

1. Akram, S.T., M.T
2. Amir Zaki Mubarak, S.T., M.T
3. Ir. Asbar, M.T
4. Ir. Darwin Harun, M.T
5. Ir. Masri Ali, M.Eng
6. Ir. Masri, M.T
7. Muhammad Tadjuddin, S.T., M.Eng.Sc
8. Sabri, S.T., M.T
9. Ir. Suhaeri, M.Eng
10. Syahriza, S.T., M.Eng
11. Teuku Firsa, S.T, M.Eng.Sc
12. Ir. Udink Aulia, M.Eng
13. Zulfadhli, S.T., M.T

IOP Conf. Series: Materials Science and Engineering **931** (2020) 011001 doi:10.1088/1757-899X/931/1/011001

IOP Conf. Series: Materials Science and Engineering **931** (2020) 011001 doi:10.1088/1757-899X/931/1/011001

10. Dr. Sarwo Edhy S, S.T., M.Eng
11. Dr. Ir. Sulaiman Thalib, M.T
12. Dr. Syarizal Fonna, S.T., M.Sc
13. Dr. Zahrul Fuadi, S.T., M.Sc
14. Dr. Ir. Hamdani, M.T
15. Dr. Muhammad Ilham Maulana, S.T., M.T

Administrators

1. Milda Laura, SE
2. Fitri Yanti, SE

IOP Conf. Series: Materials Science and Engineering **931** (2020) 011001 doi:10.1088/1757-899X/931/1/011001

About Unsyiah

Universitas Syiah Kuala (Unsyiah) is the oldest public university in Aceh. Standing on September 2, 1961 by Decree of the Minister of Higher Education and Science Number 11 of 1961, dated July 21, 1961. The establishment of Unsyiah confirmed by Decree of the President of Republic Indonesia, number 161 of 1962, dated April 24, 1962 in Kopelma Darussalam, Banda Aceh. Unsyiah domiciled in the capital of Aceh province with the main campus located in the city of Students (Kopelma) Darussalam, Banda Aceh. Currently, Unsyiah has more than 30,000 students studying in 12 faculties and the Graduate Program.

About AIC

The Annual International Conference (AIC) is not only to share the scientific knowledge but also an important to open the network among the participant and also an open-door event to promote Unsyiah for the researchers in national and International level. The 9th AIC 2019 provide an excellent forum for sharing knowledge and research findings among academicians, professionals, and governments. Topics of interests cover all theoretical and practical aspects of Sciences and Engineering, Health and Life Sciences, and Social Sciences. The primary goal AIC 2019 is organized to promote multidisciplinary understanding of the latest research and innovations through discussion, exchange, and sharing among researcher, practitioners, academicians and students.

About ICECME

International Conference on Experimental and Computational Mechanic in Engineering is intended as an International Forum for those who wish to present their latest research results, innovative ideas, and experiences in the fields of Mechanical Engineering. It also offers the excellent opportunity to meet experts, exchanges information, and strengthen the collaboration among researchers, engineers, and scholars from academia, government, and industry, while enjoying the cultural heritage and natural beauty of Banda Aceh, coupled with the friendliness of its people.

Among the topics discussed in the ICECME 2019 were (but not limited to) Material Engineering, Mechanical Engineering, and Manufacturing Processing and System. ICECME has been conducted as annual international conference and will be continued in the years on wards. The organizer has been working seriously to manage the quality of the conference, as well as to increase the number of participant and the availability of the papers in a higher level of international publication.

IOP Conf. Series: Materials Science and Engineering **931** (2020) 011001 doi:10.1088/1757-899X/931/1/011001

WELCOME SPEECH FROM RECTOR OF UNIVERSITAS SYIAH KUALA

Assalamualaikum Wa Rahmatullahi Wa Barakatuh,

In the name of Allah, the Most Beneficent and the Most Merciful. May peace, mercy, and blessings of Allah be upon you.

Dear colleagues, professors, lecturers, researchers, ladies and gentlemen. On behalf of Universitas Syiah Kuala, I would like to express my sincere gratitude and welcome you to the 9th Annual International Conference (AIC) 2019. The AIC 2019 is held in conjunction with the 2nd English Education International Conference (EEIC), the 2nd Aceh International Symposium on Civil Engineering (AISCE), and the 1st International Conference on Experimental and Computational Mechanic in Engineering (ICECME).

Moreover, I honorably welcome our keynote speakers Prof. Moses Tade from Curtin University, Australia, Dr.-Ing. Joewono Prasetijo from Universiti Tun Hussein Onn, Malaysia, Prof. Diana Elizabeth Waturangi from Atma Jaya Catholic University, Indonesia, and Assoc. Prof. Dr. Rob Amery from University of Adelaide, Australia. We also have with us invited speakers Dr. Luciana from Universitas Katolik Indonesia Atma Jaya, Indonesia, and Dr. Is Helianti from Pusat Teknologi Bioindustri, Indonesia.

It is my hope that the 9th AIC 2019 would be able to achieve its objective in providing an effective forum for academician, researchers, and practitioners to strengthen academics and industrial collaborations to enhance research and innovation. We believe that academics and industry must not work alone but must collaborate. Because with collaboration, the goals become easier, more effective, and efficient to achieve. And, we can change to a better environment and life.

We also believe knowledge, research and technology are for all people, for minorities, people with disabilities, and to help anyone. No matter how much we can accomplish by ourselves, whether it be research or development, it is never sufficient in this world of knowledge. Therefore, the focal drive of this conference is to exchange ideas, and by participating in this exchange, it is hoped that all parties who may benefit from the conference can apply it in managing activities in their areas. It is pleasing to note that the agenda of this conference covers a wide range of interesting topics related to all theoretical and practical aspects. There are Sciences and Engineering, Social Sciences, and Health and Life Sciences.

Last but not least, my deepest gratitude goes to the Advisory Board, Organizing Committee, International Scientific Committee, institutions, companies, and volunteers who have directly and indirectly supported the success of this seminar. The committee has organized a vibrant scientific program and is working hard to present highly respected and internationally notorious speakers to lead it. Although we try our finest to be professional, on behalf of Syiah Kuala University, please accept our sincere apologies should there be inconveniences that occur before, during, or after the event. I wish you a very productive conference with exciting and encouraging discussions and exchange of knowledge so that together we can anticipate a future by strengthening academics and industrial collaborations to enhance research and innovation.

May God bless us all with good health to make this event a successful and enjoyable one!

Best Regards,

PROF. DR. IR. SAMSUL RIZAL, M.ENG
Rector of Universitas Syiah Kuala

IOP Conf. Series: Materials Science and Engineering **931** (2020) 011001 doi:10.1088/1757-899X/931/1/011001

WELCOME SPEECH FROM THE GENERAL CHAIR OF THE 9TH AIC 2019

Assalamualaikum Wa Rahmatullahi Wa Barakatuh,
On behalf of the organizing committees, I would like to welcome all of you to Banda Aceh, Indonesia for the 9th Annual International Conference (AIC) 2019, Universitas Syiah Kuala. The AIC is an annual international conference carried out regularly by the university since 2011. This year, we organize the AIC in conjunction with the 2nd English Education International Conference, the 2nd Aceh International Symposium on Civil Engineering 2019 and the 1st International Conference on Experimental and Computational Mechanic in Engineering.

Under the theme of the conference "**strengthening academics and industrial collaborations to enhance research and innovation**", this conference features rich programs, including Universitas Syiah Kuala Innovation Expo 2019, four keynote speeches delivered by Prof. Moses Tade from Curtin University, Australia, Dr.- Ing. Joewono Prasetijo from Universiti Tun Hussein Onn, Malaysia, Prof. Diana Elizabeth Waturangi from Atma Jaya Catholic University, Indonesia, and Assoc. Prof. Dr. Rob Amery from University of Adelaide, Australia. Our invited speakers this year are Dr. Luciana from Universitas Katolik Indonesia Atma Jaya, Indonesia, and Dr. Is Helianti from Pusat Teknologi Bioindustri, Indonesia. The 9th AIC 2019 offers a special opportunity to bring together professors, researchers and scholars around the globe, and serves as a platform to deliver innovative research results and latest trends and development in the fields of sciences and engineering, social sciences, and health and life sciences.

All accepted papers in this conference have the opportunities to be published in various reputable publications such as IOP Material Science and Engineering Proceedings, Key Engineering Materials, Journal of Serangga, Microbiology Indonesia, Humanities and Social Sciences Reviews, International Journal of Language Studies, among others. The conference has received more than 400 submitted papers, whereby 330 papers have been accepted by the committees for presentation and to be included in the proceedings or selected for journals. To all members of the organizing committees, the international scientific committee, the reviewers, and the collaboration partners, we would like to thank all of them for their tremendous efforts to organize this conference successfully.

We look forward to having a successful conference, and we hope that all the attendees enjoy and benefit from this conference.
Best Regards,

DR.-ING. RUDI KURNIAWAN
General Chair of the 9th AIC 2019

IOP Conf. Series: Materials Science and Engineering **931** (2020) 011001 doi:10.1088/1757-899X/931/1/011001

Chairman Address

It is with a great honor to present the International Conference on Experimental and Computational Mechanics in Engineering (ICECME) 2019, held in Banda Aceh, Indonesia on 18 and 19 September 2019 by Department of Mechanical and Industrial Engineering (JTMI), Faculty of Engineering, Universitas Syiah Kuala, Banda Aceh, Indonesia. The organizer would like to appreciate Universitas Syiah Kuala for their financial support to the conference as well as facilities, professors, lecturers, admin staff and students. Special thank also extended to PT Perusahaan Listrik Negara (PLN) (Persero) Unit Induk Wilayah Aceh for their complimentary conference kit given to all conference participants.

Although it is a first event of ICECME, JTMI Department has previously conducted series of similar conferences named Computational Mechanics and Numerical Analysis (CMNA). The first Regional Seminar of CMNA was held on 2002 in Banda Aceh. The Seminars were then annually held as the 2nd, the 3rd, the 4th and the 5th one from 2003 to 2006 in Banda Aceh. In 2008 the 6th CMNA become International Conference, jointly organized with Department of Mechanical and Material Engineering, Universiti Kebangsaan Malaysia. The collaboration continued to the 7th International Conference of CMNA held in Banda Aceh on 2009. More than 200 papers from varies institutions and countries have been published in proceeding of CMNA.

The ICECME 2019 is conducted to continue the spirit of research collaboration, international publication, network meeting and industrial linkages. Sixty five papers have been accepted to be presented in the conference. The papers come from 20 institutions and 5 countries (Indonesia, Brunei Darussalam, Malaysia and Australia). After further reviewed the papers will be published in Key Engineering Materials Journal or IOP Conference series Material Science and Engineering based on topics and journal qualifications.

Finally, the organizer would like to congratulate all of the authors of the papers. Without your contribution, the success of the conference could not be possible. Hopefully, the authors and participant could share their knowledge and experiences in research and industrial application and will come-up with new network and collaboration.

Dr. Mohd. Iqbal

Chairman of ICECME 2019

IOP Conf. Series: Materials Science and Engineering **931** (2020) 011001 doi:10.1088/1757-899X/931/1/011001

Keynotes Address

Prof. Dr-Ing. Yulfian Aminanda, C-Eng (IMeche-UK)

Deputy Dean Research, Faculty of Engineering UTB Brunei Darussalam
Senate Member of UTB

Title : Finite element modeling at phenomenological / meso-structure level for industry application

The need of computation analysis on a complex and significant size of structure becomes a necessity in the era of industry 4.0. Effectively heavy industry such as aerospace, naval, automotive, civil and construction company used extensively the computation analysis especially the finite Element Analysis. The FEA is employed due to, prior to its model validation, reducing the test cost, enabling behavior prediction of structural behavior and strength, optimizing the outputs based on parameters study and hence as a tool design for certain application. The FEA becomes a necessity tool in improving the output products through an extensive research activities within the product development division of a company.

The question remains on the how detail certain physical phenomena would like to be simulated. More detail, or more elements in FEA language, will cost more on the CPU time hence cost of computation at the end. In the other hand, new proposed FE model should be able to solve a complex and big structure as a primary request for industry application. A balance of finding an appropriate FE model with its use for industrial application becomes a center of discussion among industry and research institution players.

One of the solution to obtain an FE simulation applicable for industrial use, is to propose a model at phenomenological level or meso-structure level. It consists to construct a FE model to simulate only the outputs required by users (industry) using simplest model possible. In general, the simple model could propose by introducing a new mathematical behavior or properties to the material or element which avoid to create a small size of mesh hence reduce the CPU time. Also, it can pass the modeling at micro-structure level.

The mathematical model can be obtained through coupling experiment-simulation study. A simple specimen can be performed as reference for validation, and validated FE model can be scrutinized to observed detail behavior of specimen and to propose its meso-level behavior. The meso-level behavior will be tested experimentally as well to validate the assumptions.

Two samples of research work required by The Innovation Works AirBus France can be a good example to illustrate the meso-structure model. The first one, is on the sandwich/composite structure strength subjected to impact loading. The second one, on the spring-back of CFRP laminate produced by autoclave manufacturing process. Both projects required a simple FE model in order to use it as a design tool for real aircraft structure size and its complex shape.

The proposed simple FE model for both models were obtained through experimental-FE simulation study. For the first project, it is proposed to put user material properties to the nomex hexagonal core of sandwich and model it as an array of non-linear spring. The user material properties were obtained from a simple test of uniform compression and indentation on nomex core structure. Since the spring element is used, the CPU time reduce significantly compared to model modeling the core with a small size of shell elements. For the project of spring-back, it is proposed to modify the thermal expansion coefficient of the 1st ply of CFRP in contact with mold. The modified properties are needed to simulate the shear stress produced at the 1st ply-mold interface due to different thermal expansion of 1st ply and mold. The shear stress at the interface once it reach its yield, it will be transferred to the remaining plies hence the spring-back occurs. The proposed mathematical equation of thermal expansion in function of dimension and thickness of laminate is obtained using simple curing of CFRP specimens.

Both proposed FE model can be applied to simulate a real and complex form of aircraft parts such passenger floor, wing ribs, flaps, wing spar, leading edge among others .

Brief Yulfian Aminanda CV

Yulfian Aminanda has started his career as Engineer in aerospace industry since 1993. He served as Engineer, senior engineer and supervisor of Head Department in Indonesian Aerospace Industry from 1993 – 2005 dealing with mainly aircraft structure design and analysis at the beginning of his career and end up as Head of propulsion integration system department for N2130 turbofan aircraft. In the meantime, He has worked as specialist engineer with The Boeing Company in Seattle USA dealing with certification process of B747 extended range, propulsion integration design of new generation of B737 and B777 for 2 years (1997-1998).

Starting 2005 until 2018, Yulfian Aminanda serves as Lecturer and researcher in Department of Mechanical Engineering for aerospace program of IIUM Malaysia. He started as assistant Professor in 2004 and becomes Professor in the year of 2012. He holds administrative posts since joining the faculty of engineering IIUM from Academic Advisor for Aerospace Program (2005 – 2010), coordinator for testing and industry link (2011), coordinator for training and executive program (2012) to Head of Research of the faculty and technical manager and signatory for Universal Testing Machine accreditation at the same university.

IOP Conf. Series: Materials Science and Engineering **931** (2020) 011001 doi:10.1088/1757-899X/931/1/011001

Dr Yulfian Aminanda obtained several grants from ministry of education of Malaysia in the area of composite, sandwich structure, impact engineering, finite element and others. AMIC (Aerospace Malaysian Innovation Centre) a government research agency which is grouping

industries (Airbus, Roll-Royces, CTRM etc), government agencies (MIGHT and others) and universities has granted his team a project on minimizing the reject part of aircraft composite structure produced by CTRM in the year 2013. He is also a member of Dean Council for Aerospace program under purview of Ministry of Education of Malaysia since 2005. His international joint research relates on simulation of composite manufacturing process co-organized with Institute Clement Ader, INSA and Innovation Works Airbus in France. Other research in the area of impact on composite has been completed together with ISAT, University of Burgundy in France. A study of impact on sandwich structure, a joint-research with Innovation work AirBus, has been completed as well.

Prof Yulfian has been invited as Invited Professor in ISAT, University Burgundy, Nevers France to conduct UG courses and postgraduates research supervision. And to add his international recognition, staffs mobility to enhance joint-research has been performed between him and his team in ICA Toulouse, ISAT Nevers and IIUM under sponsorship of French Embassy and AMIC.

His research works have been published in ISI, Scopus and major journals in France. A total of 65 paper publication has been registered with a total citation of 481. He was appointed to review papers on major publisher such as Applied Composite journal, Composite structure part A and B, Impact engineering.

Currently, Dr Yulfian Aminanda, is a Professor at Mechanical Engineering Program Area, Deputy Dean Research of Faculty of Engineering UTB Brunei Darussalaam and senate member at the same university.

IOP Conf. Series: Materials Science and Engineering **931** (2020) 011002 doi:10.1088/1757-899X/931/1/011002

Peer review declaration

All papers published in this volume of *IOP Conference Series: Materials Science and Engineering* have been peer reviewed through processes administered by the Editors. Reviews were conducted by expert referees to the professional and scientific standards expected of a proceedings journal published by IOP Publishing.

- **Type of peer review:** Single-blind
- **Conference** submission management system: Email process, managed by (akhyar@unsyiah.ac.id)
- **Number of submissions received:** 56
- **Number of submissions sent for review:** 42
- **Number of submissions accepted:** 24
- **Acceptance Rate (Number of Submissions Accepted / Number of Submissions Received X 100):** 57.14 %
- **Average number of reviews per paper:** 3-4 manuscripts
- **Total number of reviewers involved:** 12 reviewers
- **Any additional info on review process:** Papers/manuscript after revised by authors following by reviewers comments, than final manuscript was proofread by professional proofreading, it is English Language Center of Universitas Syiah Kuala - Indonesia.
- **Contact person for queries:** Mohd. Iqbal (mohd.iqbal@unsyiah.ac.id) and Akhyar (akhyar@unsyiah.ac.id)

IOP Conf. Series: Materials Science and Engineering **931** (2020) 012001 doi:10.1088/1757-899X/931/1/012001

The Sensitivity Analysis in Topology Optimization of Hip Stem Prosthesis Using Finite Element Method

SYIFAUL Huzni[1], M. ILHAM Tanamas[1], SYARIZAL Fonna[1] and A.K. ARIFFIN[2]

[1]Dept. of Mechanical and Industrial Engineering, Faculty of Engineering, Universitas Syiah Kuala, Banda Aceh, Indonesia

[2]Centre for Integrated Design for Advanced Mechanical Systems (PRISMA), Universiti Kebangsaan Malaysia

E-mail: syifaul@unsyiah.ac.id

Abstract. This paper discusses the results of the parameters employed sensitivity analysis in the optimization of hip stem prosthesis geometry using topology optimization techniques based on finite element methods. The study was carried out to examine the relationship of finite element parameters, such as mesh size and stress distribution, and topology optimization parameters, such as percent to retain, to the final geometry of the prosthesis stem. Topology optimization was performed using ANSYS software for hip stem prosthesis, a type of Anatomic Medullary Locking, made from titanium alloy. The results obtained from this study indicate that the selection of finite element parameters and topology optimization parameters will have a significant effect on the shape and geometry of the hip stem. The results of this analysis can be the basis for the development of hip stem shape and geometry through topology optimization techniques.

1. Introduction

Topology optimization based on the finite element analysis is one of the optimization techniques that is often employed in the process of shape optimization. This technique has been implemented in various fields of science, such as automotive, structural engineering, aerospace, and biomechanics, to produce a more optimal design [1]. Topology optimization works by taking a solid block of material in any shape and removes material from it to minimize or maximize an optimization objective such as mass, displacement, or stress.

Numerically, the process of analysis and optimization involves several parameters that make up the model used in the analysis. Variations in the value of the parameters used may cause differences in the results obtained. To reduce the effect of uncertainty on the optimization results that will be achieved when optimization is carried out, it is necessary to analyze the sensitivity of the parameters used in finite element analysis and also the parameters used in topology optimization. Therefore, in this paper, we will discuss the results of investigations on the two parameters employed in the hip-stem-prosthesis optimization process, i.e., mesh size and percent of retain

IOP Conf. Series: Materials Science and Engineering **931** (2020) 012001 doi:10.1088/1757-899X/931/1/012001

2. Geometry and Material

The dimensions of the prosthesis used in this study refer to those developed by previous researchers, namely based on anthropometric data of the human body of Indonesian people aged between 22-50 years old. Previous researcher has conducted an analysis related to the safety level of hip stem prosthesis, Anatomic Medullary Locking type, using geometry adjusted for Indonesian people's body posture, with a stem length of 140 mm, a ball head diameter of 38 mm and an extrusion thickness of 15 mm [2]. Detailed dimensions of the model used in this study are shown in Fig. 1. The boundary conditions given to the prosthesis for finite element analysis are shown in Fig. 2, with the force component for standing conditions of Fx = 0% BW; Fy = -100% BW; Fz = 0% BW, where the body weight (BW) selected is 70 kg [3,4]. Material properties used in finite element analysis for hip stem prosthesis refer to Ti-6Al-4V titanium alloy material, with a poisson ratio (v) = 0.342, modulus of elasticity (E) = 113.8 GPa [5].

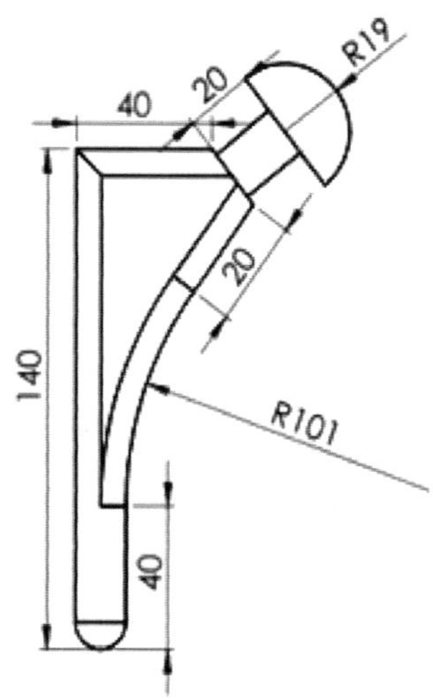

Figure 1. Geometry of prosthesis

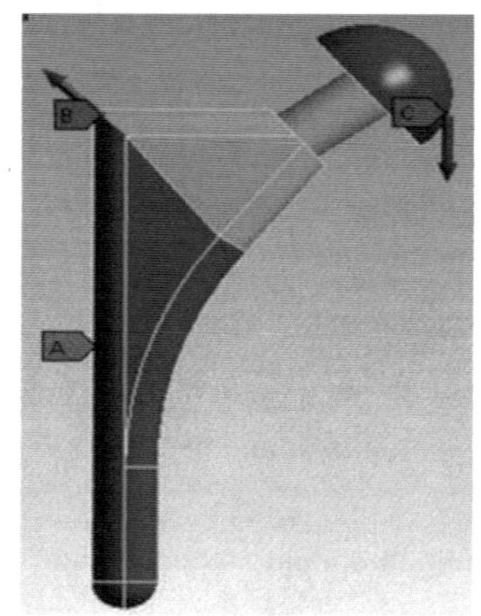

Figure 2. Boundary Condition

3. Control parameters

Finite element Analysis. The type of mesh employed in this study is tetrahedrons. Three variations of the mesh size used are fine, medium and coarse, according to the default conditions in ANSYS software. When examined in more detail, the average mesh size for fine is 3 mm, medium is 5 mm and coarse is 10 mm. The arrangement of the mesh in the hip stem prosthesis based on its size can be seen in Fig. 3.

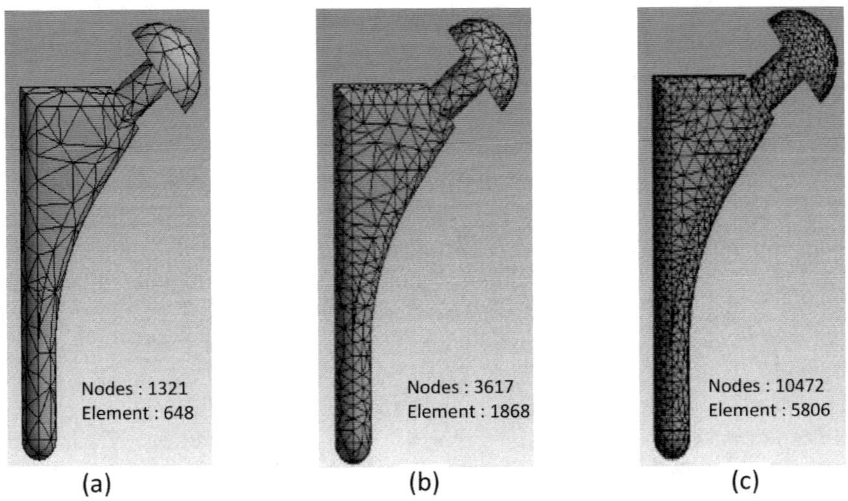

Figure 3. Meshing (a) Coarse, (b) Medium, (c) Fine

Topology Optimization. In the topology optimization procedure, one important control parameter is percent-of-retained (POR). POR is used as a determinant of the area that can be optimized with the topology method. In this study, three POR values were chosen namely 40%, 60% and 80%, with minimizing compliance as an objective. Determination of the area removed is based on the distribution of stress magnitude from the finite element analysis results.

4. Result and Discussion

Correlation between mesh size and stress distribution. The results of finite element analysis to study the effect of mesh size on the stress distribution of the prosthesis used for standing conditions are shown in Fig. 4. The highest concentration of von-mises stress is in the neck of the hip stem prosthesis. The maximum value of von-mises stress for mesh coarse size is 41,136 MPa, medium 47,424 MPa, and fine 50,609 MPa. The differences occur due to the differences in the mesh size used in finite element analysis, as can be seen in Fig. 4 and Fig. 5.

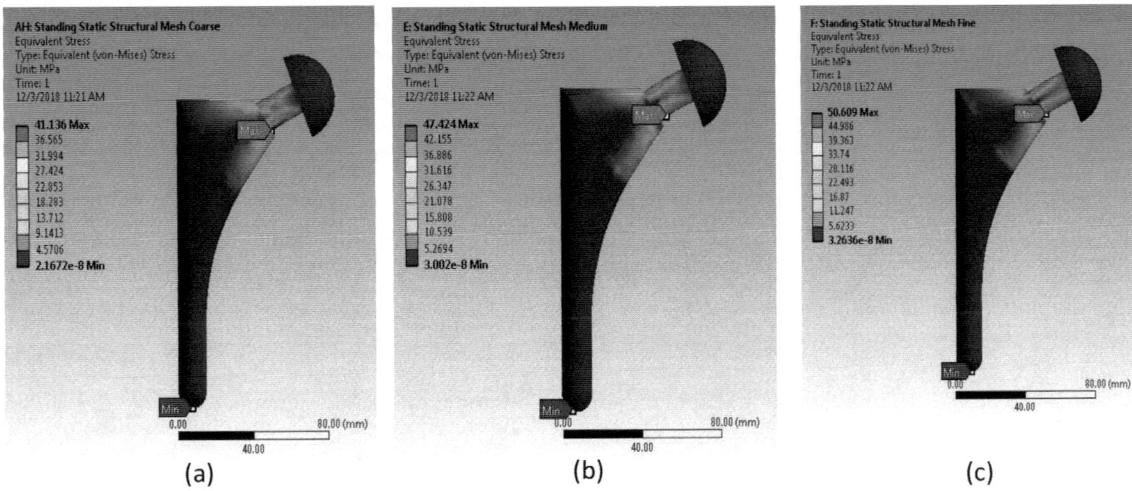

Figure 4. Von Misses stress distribution of pre optimization prosthesis a) Coarse, (b) Medium, (c) Fine

IOP Conf. Series: Materials Science and Engineering **931** (2020) 012001 doi:10.1088/1757-899X/931/1/012001

The relationship between mesh size and optimization results. The simulation results that have been carried out show that mesh size influences the final form of optimization results with topology techniques for hip stem prosthesis. The results of topology optimization show that removing only occurs in part of the predetermined design region. The process of removing that occurs is based on the level of stress that occurs. Parts of the prosthesis model that experience minimum stress will be removed element by element [6]. The size of the removed model depends on what percentage to retain (POR), and the size of the mesh used also has an effect on the removal process as shown in Fig. 6.

Stress Analysis of the optimized model. The simulation results show that there are differences in stress distribution between models that have been optimized using topology optimization techniques. Nevertheless, the stress distribution on the prosthesis shows an even distribution, not only focusing on a certain area. The maximum von-Mises stress value for the prosthesis model with 80% to retain is 87,836 MPa, 60% to retain is 76,257 MPa, and 40% to retain is 126.9 MPa.

Prosthesis with 40% to retain shows the highest maximum von-mises stress magnitude, compared to the model with 60% or 80% to retain as shown in Fig. 7. In this case, the smaller percent to retain is used, the von Mises stress distribution will be more evenly distributed, and no longer centered on the neck of the hip stem prosthesis as found in prosthesis that has not been optimized. Even so, the maximum value of von Misses stress occurred is still below the value of material yield, so it can still be considered safe to apply.

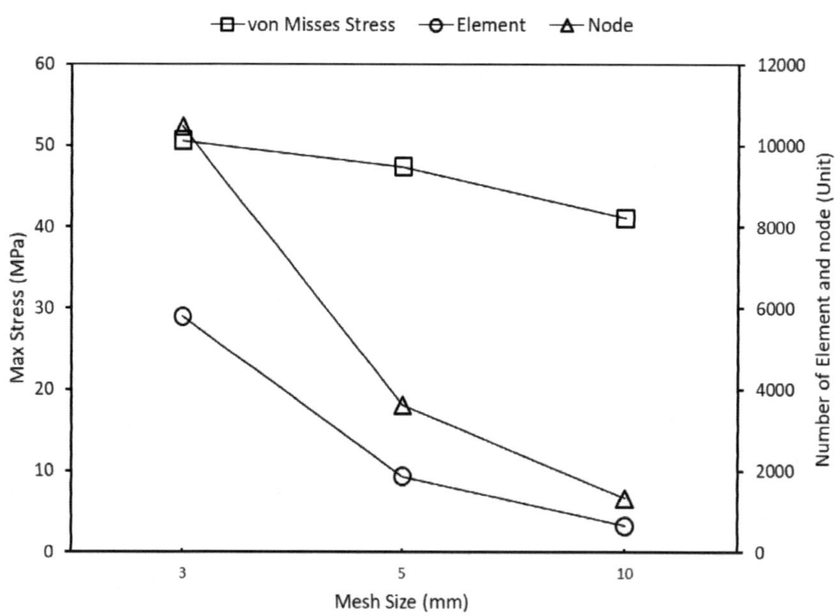

Figure 5. Corelation between mesh size, number of element and maximum stress

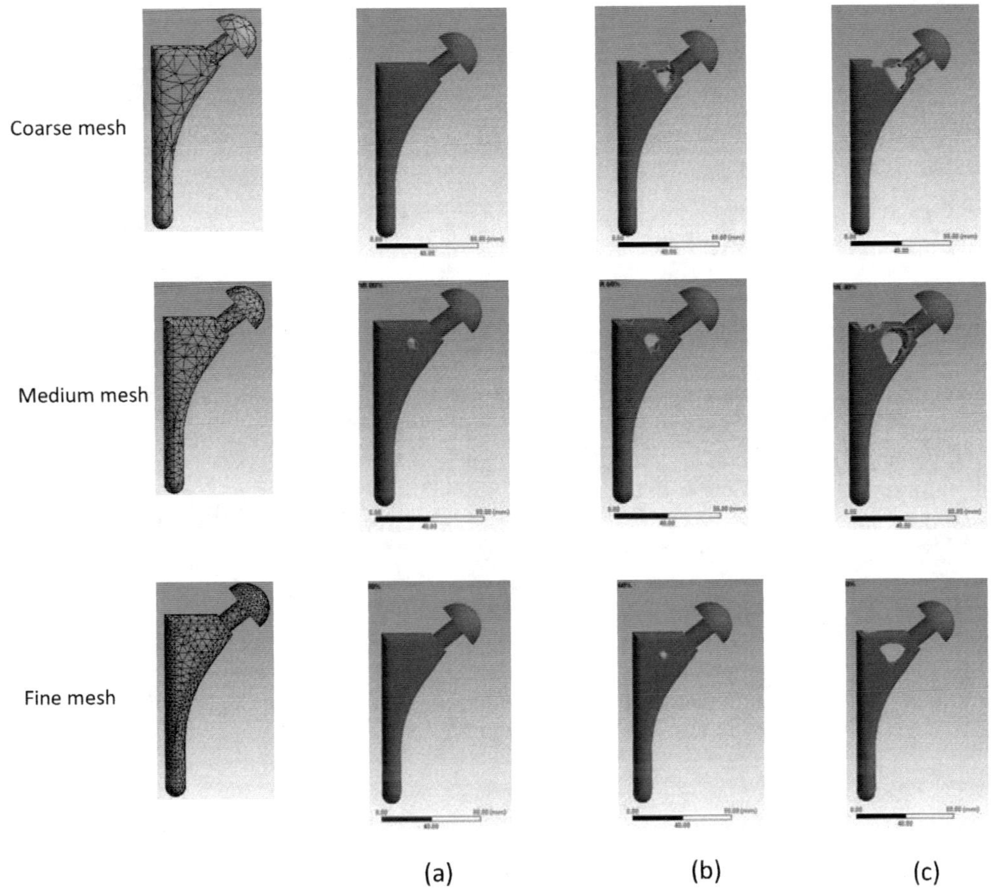

Figure 6. Mesh vs. Retain Percentage (a) 80%, (b) 60%, (c) 40%

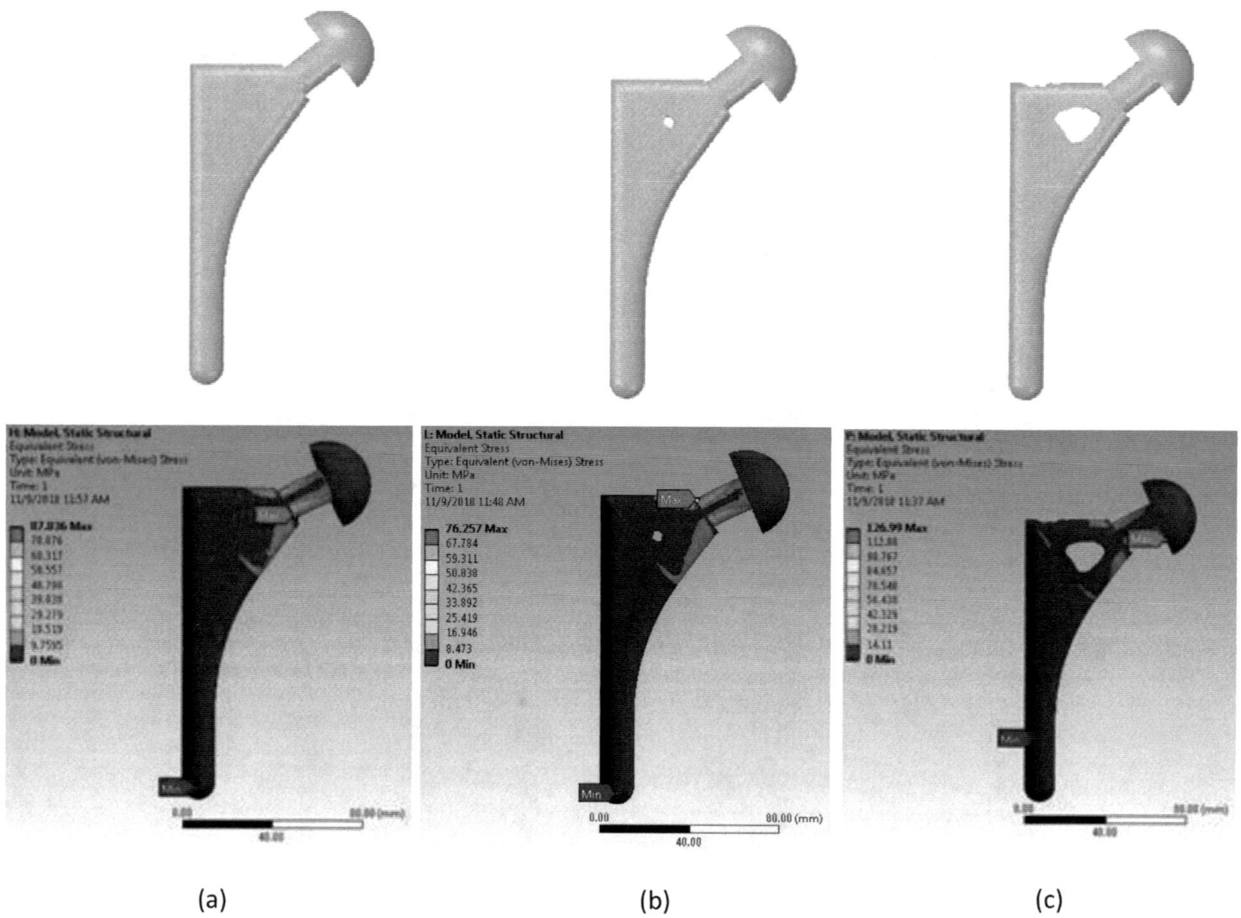

Figure 7. Von Misses stress distribution of post optimization prosthesis on fine mesh (a) 80%, (b) 60%, (c) 40%

5. Summary

The results of sensitivity analysis that have been carried out in this study show that the mesh size, which is a finite element analysis parameter, and the retain percentage, which is a parameter of topology optimization, influences the final shape and geometry of the optimization results on the hip stem prosthesis. Fine mesh gives the analysis results of a more sensitive Fine mesh that provides the results of stress analysis with better sensitivity, this can be seen from the magnitude of stress that is greater than other mesh sizes. Mesh size also influences on the shape and final geometry of the optimization of the hip stem which is done by topology techniques.

Acknowledgement.

The research was supported by Penelitian Lektor Sumber Dana PNBP Universitas Syiah Kuala 2019 with contract number of 224/UN11.2/PP/PNBP/SP3/2019.

IOP Conf. Series: Materials Science and Engineering **931** (2020) 012001 doi:10.1088/1757-899X/931/1/012001

References

[1] Martin P. Bendsøe & Ole Sigmund, *Topology Optimization: Theory, Methods, and Applications*, Springer-Verlag, New York, 2004

[2] Syifaul Huzni, Fitri Handayani, Syarizal Fonna & M. Ridha, *Analisis Hubungan Dimensi Stem dan Ballhead Terhadap Faktor Keamanan Fatik Hip Stem Prosthesis dengan Menggunakan Metode Elemen Hingga*, Prosiding Seminar Nasional Tahunan Teknik Mesin XIV, Program Studi Teknik Mesin Universitas Lambung Mangkurat (2015)

[3] Syifaul Huzni, Rizky S, Syarizal Fonna, Sulaiman Thalib, M. Ridha, A.K Arifin, Analisa *Distribusi Tegangan pada Hip Stem Prosthesis dengan Menggunakan Metode Elemen Hingga*, Proseding Seminar Nasional Rekayasa (SNTR) III Tahun 2016, Politeknik Aceh Selatan (2016).

[4] *Loading of Orthopedic Implants*, information on www.otrhoload.com/database/

[5] *Titanium Ti-6Al-4V (Grade 5) Annealed*, information on http://asm.matweb.com/search/SpecificMaterial.asp?bassnum=mtp641

[6] Picelli, R., Townsed, S., et al., *Stress-based Shape and Topology Optimization with the Level Set Method*, J Computer Methods in Applied Mechanics and Engineering. (2017) pp. 1-23.

IOP Conf. Series: Materials Science and Engineering **931** (2020) 012002 doi:10.1088/1757-899X/931/1/012002

Step by Step Stress Analysis of Ammonia Scrubber Vertical Pressure Vessel Using Finite Element Method

Arhami, Amir Zaki Mubarak, Andika Saputra, Masri Ali

Department of Mechanical and Industrial Engineering, Faculty of Engineering, Universitas Syiah Kuala, Banda Aceh, Indonesia

E-mail: arhami@unsyiah.ac.id

Abstract. In order to solve the stress problem of ammonia scrubber pressure vessel, this paper proposes the analysis steps using finite element method. The study is a type of vertical pressure vessel used to remove pollutants in ammonia gas. Firstly, the modeling of the vertical pressure vessel geometry was carried out using Autodesk Inventor Professional 2017. Secondly, the meshing was carried out using Autodesk Nastran In-CAD 2017. For the meshing, we used tetrahedral element and we chose the parabolic type element order. Thirdly, the analysis was carried out using Autodesk Nastran In-CAD and based on the design pressure of 0,5 MPa with a volume of 1,21 m^3. The results showed that the maximum stress occurred in the connection area between the inspection opening and the shell of 333,4 MPa. The maximum displacement occurred in the shell adjacent to the inspection opening was 3,268 mm. Based on ASTM standards on SA 240/A240M Gr 304 material, the maximum strength of material stress was 515 MPa. Finally, theoretical validation was carried out for the entire model, and the results were within the permissible limit.

1. Introduction

Based on field conditions, popping gas from the low-pressure system could not be handled by the existing ammonia scrubber because the gas mixture entering the existing ammonia scrubber is already too much, so the loaded gas that must be absorbed has exceeded the capacity. To overcome this problem, it was planned to install a special ammonia scrubber for gas output from the low-pressure system so that the exhaust gas is not exposed to the environment [1].

Ammonia scrubber in its design requires high accuracy because the ammonia is a dangerous substance. The author will design and analyze the stress occurred in the ammonia scrubber pressure vessels. Stress in a pressure vessel cannot be ignored since it is the basis for determining the dimensions and the type of material used in a pressure vessel.

In the designed step of pressure vessels, the most important thing is to take into account the stresses that appear on the walls of the pressure vessels. Stresses that occur can be caused by external factors such as disaster effect of wind and earthquake loads as well as internal factors caused by work pressure instability and the weight of the vessel itself. The modeling was completed on a modeling software Autodesk Inventor Professional, and a finite element analysis assisted by Autodesk Nastran In-CAD software was carried out to highlight the stress concentration and displacement of various points.

The pressure vessels analyzed were the vertical type with the ammonia gas fluid and assumed as a static load. The analytical application and software approaches are expected to be able to provide the stress analysis step that occurs in pressure vessels so that the magnitude and position of the maximum stress can be known , in order to avoid undesirable things [2]. To be used safely, the design results and stresses refer to ASTM standards on SA 240/A240M Gr 304 material [3]. If the stress occured is greater than the yield strength of the material used, it will cause the material failure. Theoretical validation was carried out for the entire model, and the results were within the permissible limit.

Stress condition of pressure vessel

The stresses in pressure vessels caused by operating loads are divided into two types namely if the internal pressure (P) is less than 1/6 of the permissible stress, it is called a thin-walled vessel, and if the internal pressure (P) is greater than 1/6 of the allowable stress, it is called a thick-walled vessel [4]. In this study, the pressure vessel as shown in Fig. 1 is assumed to be a thin cylinder. Therefore, the analysis follows the thin cylinder formulae.

a) Longitudinal stress

$$\sigma_L = \frac{PD}{4t} \tag{1}$$

b) Circumferential stress

$$\sigma_\phi = \frac{PD}{2t} \tag{2}$$

where t is the wall thickness in mm, P is the internal pressure in MPa, D or 2r is the inside diameter of vessel in mm, σ_L is the longitudinal stress in MPa, and σ_ϕ is the circumferential stress in MPa.

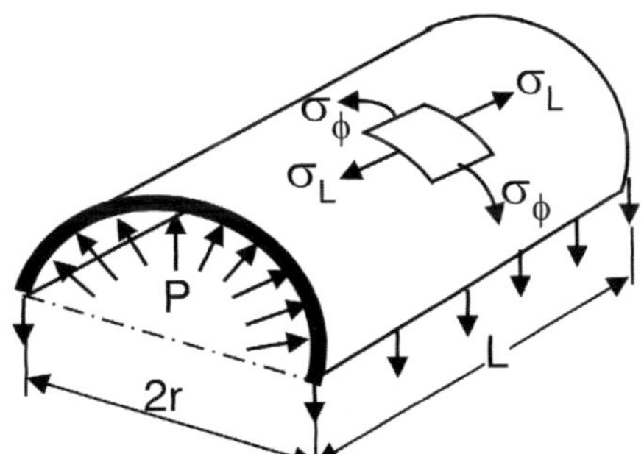

Figure 1. Stress conditions on the cylinder wall of the pressure vessel [4]

Stress condition of Head

Stresses occured on the head in several locations are acquired using the following formula,

a) At the junction of crown and knuckle area

$$\sigma_L = \frac{PL}{2t}\text{, and } \sigma_\phi = \frac{PL}{4t}\left(3 - \frac{L}{R}\right) \tag{3}$$

IOP Conf. Series: Materials Science and Engineering **931** (2020) 012002 doi:10.1088/1757-899X/931/1/012002

b) Inside of crown area

$$\sigma_L = \frac{PL}{2t}, \text{ and } \sigma_\phi = \sigma_L \quad (4)$$

c) Inside of knuckle area

$$\sigma_L = \frac{PL}{2t}, \text{ and } \sigma_\phi = \frac{PL}{t}\left(1 - \frac{L}{2r}\right) \quad (5)$$

d) At the tangent line

$$\sigma_L = \frac{PR}{2t}, \text{ and } \sigma_\phi = \frac{PR}{t} \quad (6)$$

where L is the crown radius in mm, R is radius of head in mm, and r is radius of knuckle in mm.

Equivalent Stress

Equivalent stress (equivalent) is obtained from the relationship of three-dimensional main stresses namely the longitudinal, circular, and radial directions. There are denoted as σ_L, σ_ϕ, σ_r and for equivalent stress σ', then the formula can be written as follows [5]

$$\sigma' = \sqrt{\frac{(\sigma_L - \sigma_\phi)^2 + (\sigma_\phi - \sigma_r)^2 + (\sigma_L - \sigma_r)^2}{2}} \quad (7)$$

If the stresses occurred only in circumference and longitudinal directions, the equation (7) becomes

$$\sigma' = \sqrt{\sigma_L^2 + \sigma_\phi^2 - \sigma_L \sigma_\phi} \quad (8)$$

2. Finite Element Analysis

Finite element analysis is one of the numerical methods that can be used to obtain solutions for various engineering problems. This method is widely used to calculate the structure due to the loading or other effects on the structure. The main obje`ctive is to obtain the value of the stress approach occurred in the structure. This approach is an analysis based on stress or strain assumptions [6].

Modeling Step of Vertical Pressure Vessel

In this step, the 3D geometry model of vertical pressure vessels created by the Autodesk Inventor Professional 2017. Fig. 2 shows the models made in 2D and 3D forms based on the designed results.

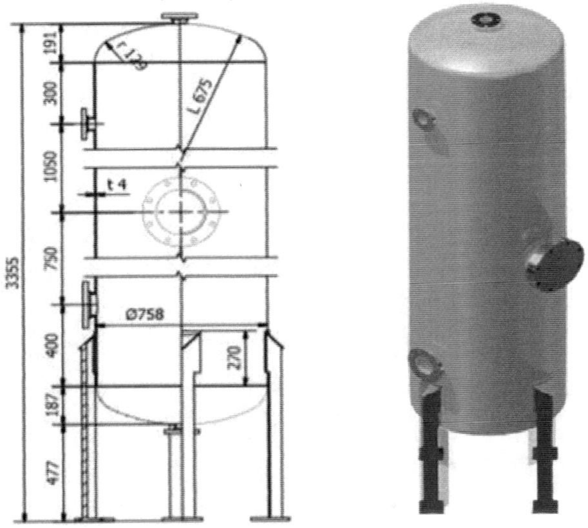

Figure 2. Vertical pressure vessel model

The proposed dimension and design specifications of the pressure vessels analyzed and material grades are listed in Table 1 and 2.

Table 1. Pressure vessel specification and design data

Fluid	Ammonia gas
Density (ρ)	0.73 kg/m^3
Operating pressure (P_o)	0.44 MPa
Design pressure (P_d)	0.5 MPa
Operating temperature (T_o)	35°C
Design temperature (T_d)	100°C
Inside diameter (D_i)	750 mm
High of shell (H)	2500 mm
High of head (h)	187.5 mm
Thickness of shell (t)	3.3 mm
Crown radius (L)	675 mm
Knuckle radius (r)	128.9 mm
Corrosion factor	0.5
Capacity	1.21 m^3

Table 2. Material specification

Component type	**Material grade**
Shell	SA 240 Gr.304
Head	SA 240 Gr.304
Nozzle	SA 182 Gr.304

Meshing Step of Vertical Pressure Vessel

At the pre-processing stage, there are several things determining the division of the elements (meshing) namely the determination of the type of material, idealization and selection of the elements. The element used to conduct stress analysis on pressure vessels is the tetrahedral type.

IOP Conf. Series: Materials Science and Engineering **931** (2020) 012002 doi:10.1088/1757-899X/931/1/012002

Meshing on the vertical pressure vessels is conducted thoroughly. The first step that must be done is to set the size of the element to 15 mm with a tolerance of 0.00272846 mm, and then choose the element order as parabolic. Fig. 3 shows the results of the division of elements (meshing) with the type of tetrahedral elements. The total numbers of nodes are 532118 and elements 269371.

Figure 3. Meshing model of equipment

Stress Analysis Step of Vertical Pressure Vessel

This step is a determination for loading on pressure vessels. The degree of freedom which becomes a constraint or boundary condition was also determined so that the reference surface is fixed. Loading in pressure vessel analysis was carried out under internal pressure uniform acting on the entire surface of the inner cylinder as shown in Fig. 4 and 5. The loading was applied to the inner pressure vessel with a pressure of 0.5 MPa.

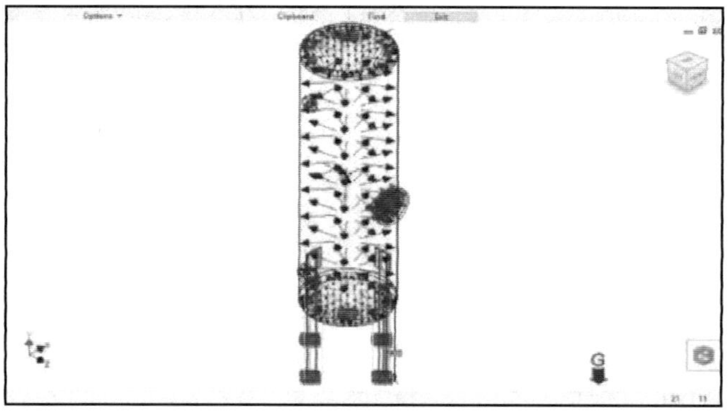

Figure 4. Application of pressure load in the pressure vessel

IOP Conf. Series: Materials Science and Engineering **931** (2020) 012002 doi:10.1088/1757-899X/931/1/012002

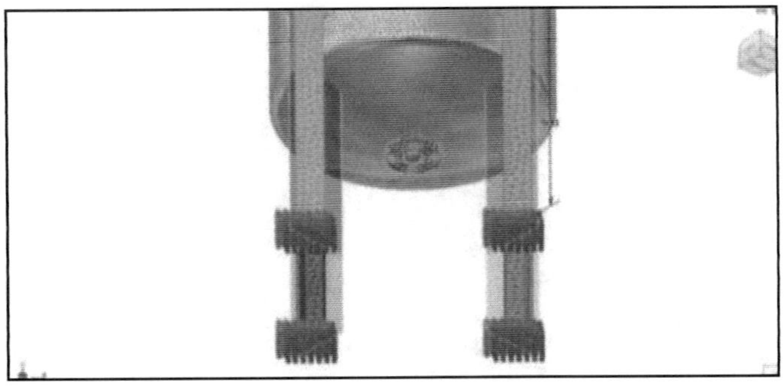

Figure 5. Determination of boundary conditions on the support

In principle, the stresses occurred in pressure vessels caused by operating loads are longitudinal stresses (σ_L) and circular stresses ($\sigma_\emptyset$). The formulation used and stress calculation results can be seen in Table 3.

Table 3. Formulation used and stress calculation results

Stress location	Equation	σ_ϕ, MPa	σ_L, MPa
Shell	(2) and (1)	56	28
Junction of crown and knuckle	(3)	30	50
Inside crown	(4)	50	50
Inside knuckle	(5)	- 162	50
Tangent line	(6)	56	28

3. Result and discussion

The results of stress distribution simulations on ammonia scrubber pressure vessels can be seen in Fig. 6 (a). The figure shows the several colors produced due to the stress conditions occured in pressure vessels. It appears that the maximum stress occured in the connection area between the inspection opening and the shell. Fig 6 (b) shows that the value of the largest stress distribution occured in the X-Y plane was at location 13 namely the curvature of the lower head where the stress value was 124.4 MPa, and the smallest one was seen at location 11 in the connection between the lower head and shell where the stress value was 25.22 MPa.

Fig. 7 (a) shows the location points determined to the Y-Z plane based on the color difference (stress value). Fig. 7 (b) shows that the value of the largest stress distribution occured in the Y-Z plane was at location 11 namely the connection area between the inspection opening and the shell with a stress value of 333.4 MPa, the smallest stress occured at location 2 precisely in the area of the peak of the upper head with stress rated of 35.5 MPa.

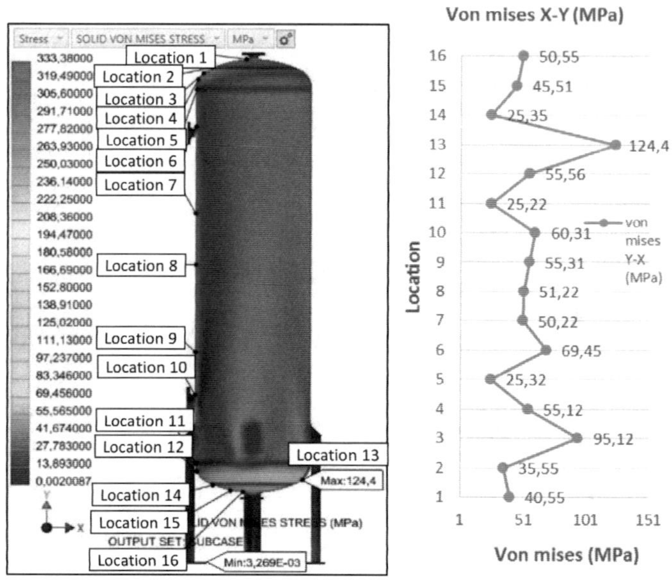

Figure 6. (a) The stress distribution location determined against the X-Y plane, (b) Graph of the stress distribution value in the X-Y plane.

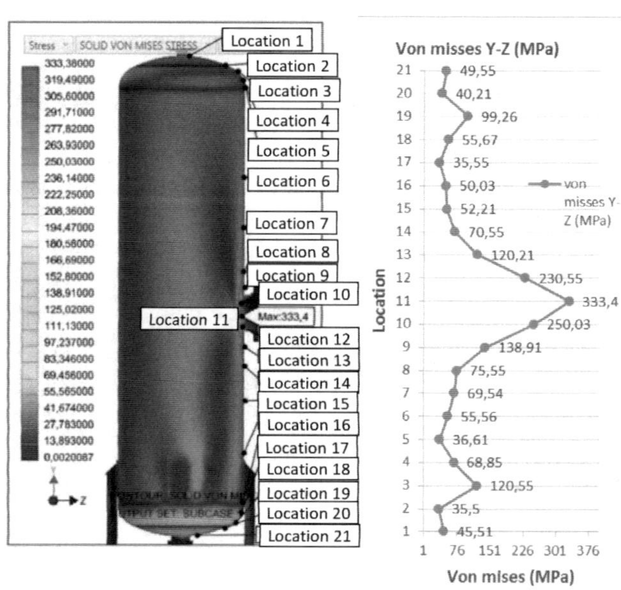

Figure 7. (a) The stress distribution location determined against the Y-Z plane, (b) Graph of the stress distribution value in the Y-Z plane

The value of displacement occurred in pressure vessels can be seen in the following figures. Fig. 8 (a) shows some location points that are determined against the Y-X plane based on the color difference occurred. To find out the amount of displacement occurred in the area can be seen in Fig. 8 (b). The figure shows the greatest displacement value occurred at location 2 precisely in the top head region with a value of 1.505 mm, displacement occurred at the connection of the head with the shell is shown by location 8 with a value of 1.21 mm.

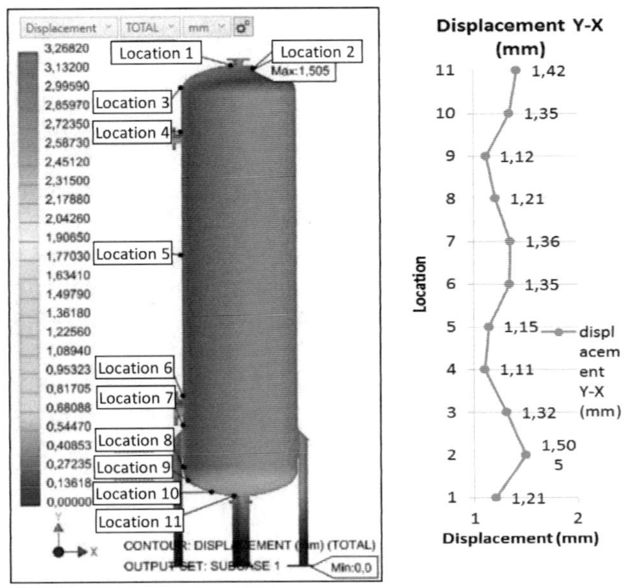

Figure 8. (a) Displacement location determined against the X-Y plane, (b) graphs of displacement values in the X-Y plane.

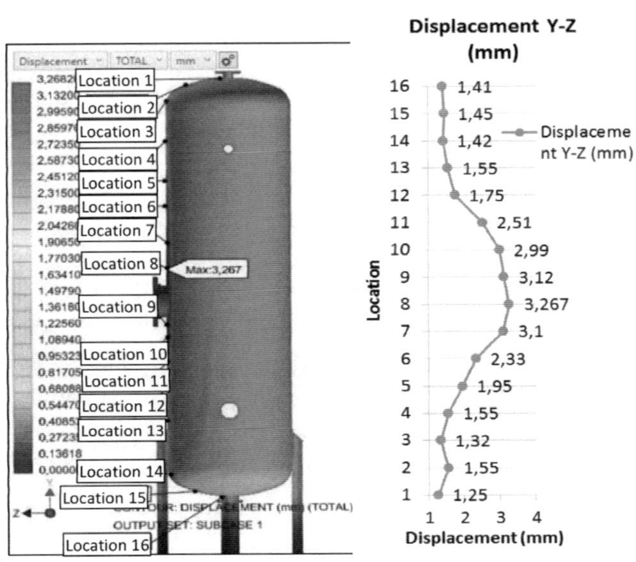

Figure 9. (a) Displacement location determined against the Z-Y plane, (b) graphs of displacement values in the Z-Y plane.

Fig. 9 (a) shows several locations that are determined against the Y-Z plane based on the color difference occured as shown in Fig. 9. To find out the amount of displacement occured in the area can be seen in Fig. 9 (b).

Fig. 9 (b) shows the maximum displacement value occured at location 8 precisely on the tube body adjacent to the inspection opening with a value of 3.267 mm, displacement occured at the curvature of the upper head is marked by location 3 with a value of 1.32 mm, while displacement occured in the lower head region is marked by location 15 with a value of 1.45 mm.

Theoretical Validation of Stress Analysis

The calculation results of the failure analysis occurred in the pressure vessel showed the greatest stress occurred on the inside of the knuckle radius shown based on an equivalent stress value of 192 MPa. This shows that the pressure vessel theories is in safe condition, since the value of the equivalent stress obtained does not exceed the value of the yield strength which is equal to 205 MPa.

The simulation results using analysis software show the maximum stress value occurred in the pressure vessel is in the connection area between the inspection opening with the shell of 333.4 MPa. Maximum displacement occurred in the shell adjacent to the inspection opening of 3,268 mm. Based on ASTM standards theories on SA 240/A240M Gr 304 material, the maximum strength of the material stress is 515 MPa. Thus, the stress occurred in the pressure vessel design simulation results is 333.4 MPa and still within the permitted limits. It can be concluded that the pressure vessel condition is safe to be used.

Summary

In this paper, the steps analysis using finite element method to solve the stress problem on ammonia scrubber pressure vessel was proposed. The study is a type of vertical pressure vessel used to remove pollutants in ammonia gas. The modelling of the vertical pressure vessel geometry was carried out using Autodesk Inventor Professional 2017 and meshing was carried out using Autodesk Nastran In-CAD 2017. For the meshing, we used tetrahedral element and element order we chose the parabolic type. The analysis was carried out using Autodesk Nastran In-CAD and based on the design pressure of 0,5 MPa with a volume of 1,21 m^3. The results showed the maximum stress value occurred in the connection area between the inspection opening and the shell of 333,4 MPa. The stress values were less then the ASTM standards on SA 240/A240M Gr 304 material, the maximum strength of material stress is 515 MPa. The maximum displacement occured in the shell adjacent to the inspection opening was 3,268 mm. Based on theoretical validationcarried out for the entire model, the results are within the permissible limit.

References

[1] Moss Dennis R., 1987. *Pressure Vessel Design Manual, Illustrated Procedures for Solving Every Major Pressure Vessel Design Problem.* Gulf Publishing Company, Houston.

[2] Manullang E., Tangkuman S., Maluegha B.L, Stress Analysis of Vertical Pressure Vessel *13ZL100040291 at Aneka Gas Industry Ltd.* Journal of Mechanical Engineering Dept. Sam Ratulangi University. (2016) 92-105

[3] ASTM, 2010, *Annual Book of ASTM Standard*, ASTM International, 100 Barr Harbor Drive, PO Box C700, West Conshohocken.

[4] Popov E. P., 1990. *Engineering Mechanics of Solid*, Prentice-Hall International, Englewood Cliffs, New Jersey.

[5] Budynas Richard G., Nisbett J.keith, 2015, Shiglhey's *Mechanical Engineering Design.* Tenth Edition, McGraw-Hill Education, New York.

[6] Moaveni, Saeed. 1999. *Finite Element Analysis Theory and Application with Ansys*, Prentice Hall, Inc. New Jersey.

IOP Conf. Series: Materials Science and Engineering **931** (2020) 012003 doi:10.1088/1757-899X/931/1/012003

Communication Support Tool With Data Glove Concept For People With Hearing Disability Using Kansai Engineering Method

Syahriza[1], S Zuhri[2], M Adilah[2]
[1]Mechaninal and Industrial Engineering Department, Universitas Syiah Kuala, Darussalam, Banda Aceh, Indonesia
[2]Industrial Engineering Department Universitas Syiah Kuala, Darussalam, Banda Aceh, Indonesia

E-mail: syahriza@unsyiah.ac.id

Abstract. The difficulty of communication that faced by the deaf and mute people make them complicated in doing activity like normal people. Sign language as a communication tool that has been formed at this time has not yet been able to connect communication with the community yet. It is only able to bridge communication among those fellow deaf and mute people. Therefore a tool which is able to translate sign language used by people with disabilities is needed. Thetool which able to translate sign language wascreated by Data Glove concept in design which can translate the word visually. Using the assistance of the Semantic Differential questionnaire, we can get the emotion level from user based on 12 Kansei words. The response in emotion form which is expresssed by users is one of theelements of the Kansei Engineering Method. Thus product adjustment to Kansei engineering is advisable and society.

1. Introduction

According to the World Health Organization [1], more than 5% of the world's population, which is 360 million people have an inability to hear sound well. Therefore, some of the people feel discriminated and hampered to socialize with social activities in general. In addition, such disability makes them often rejected from the rights to live independently, work, or even to move freely in a different social world than they are [2].

The formation of sign language as a medium of communication for the deaf and mute people at this time is really helping them interact with others. However, some people are not familiar with the use of sign language [3]. This proves that the use of sign language has not been able to connect the interaction between people with deaf-mute people to society.Thus can be the basic of the importance of making a tool that can support the deaf and mute people in communicating.

Making a product or a tool, of course, can not be separated from the method or technique that is able to support the process of designing and developing the product itself. The use of the method aims to create products that can be in accordance with the wishes of consumers themselves. In this research the method used is Kansei Engineering method. This method has been widely used, including in the development of robots for therapy and friends for elderly people (seniors) in Malaysia [4], a learning mediator robot for people with autism developed in Malaysia [5], and alarm clock made of bamboo developed in Indonesia [6]. The development of the three products using the Kansei Engineering method aims to meet the needs of consumers based on the emotions they show by using these products. The emotion shown by the user is called Kansei [7]. By using Kansei Engineering method, it will be developed a product as a translator tool that will be tailored to the user's emotions in the form of a sign language interpreter.

Technological development in the form of sign language interpreter is now developed, but the price offered is so high in result to not able to reach the market well. In addition, the design of this translator tool has also been done in Indonesia, but the tool has not been quite successful. Translation of sign language with Kinnect (microsoft products in the form of sensors) has not been able to detect the movements of the fingers, resulting in many non-translatable vocabularies [8].Translation by using Opencv software is only capable of detecting hand movements in an upright state with a distance not exceeding 90 cm [9]. Translation by using the concept of Data Glove is a concept which is able to translate sign language well by using gloves that have been embedded with sensors to detect the movement of human fingers on both hands of the user [10]. Therefore the selection of the concept of Data Glove is very appropriate to be used to detect the movement of sign language by pinning the sensor.

This concept selection is tailored to the relationship between designed productsand the needs of consumers who generally tend to use the movement of the finger in communicating compared with other limbs.Therefore the concept of Data Glove is the proper concept and it able to solve the problem of sign language translation widely without any limitations caused by certain factors. To accomplish the design of this translator product, the researcher chose Kansei engineering concept to adapt the customer's emotion to the designed produsct so that it can meet the customer's need well.

2. Research method

The initial step of this research is identifying the problem. This is preceded by looking directly at how the interaction is done by the person in the use of sign language as the primary language in order to solve the problem. This problem solving is done by collecting data in the form of sign language and Kansei Word as seen in Table 1 as secondary data of research. In addition, the primary data contained in this study is the data from questionnaires on Kansei Word filled in by respondents after the respondent using a translator tool that has been designed.

Table 1. Data of Kansei Word

No	*Kansei Word*	Definition	*Kansei Word*	*Reference*
1	Cheap	Cheaper price	Expensive	[11]
2	Simple	Simple shape and ergonomic	Complex	
3	Light weight	Has light size	Heavy	
4	Practical	Easy to use	Difficult	
5	Innovative	Innovative	Plain	
6	Rigid	Rigid and stiff	Flexible	
7	Futuristic	Modern and sophisticated	Conventional	[12]
8	Informative	Giving the impression fits data needed	Confusing	
9	Natural	Giving a natural impression	Engineered	
10	Comfortable	Brings calm and comfort feeling	Uncomfortable	
11	Enthusiastic	Gives a vibrant and passionate effect when using it	Boring	
12	*Colorful*	Full of color, using eye-catching color	Plain color	

IOP Conf. Series: Materials Science and Engineering **931** (2020) 012003 doi:10.1088/1757-899X/931/1/012003

3. Hardware design

The design of hardware is the most important thing to start the process of making this sign language interpreter. This design consists of Arduino design which acts as a master against 5 flex sensor units, DFplayer Mini, Light Computer Display (LCD), and Buzzer Speaker. Figure 1. shows the overall hardware design contained in the sign language interpreter tool.

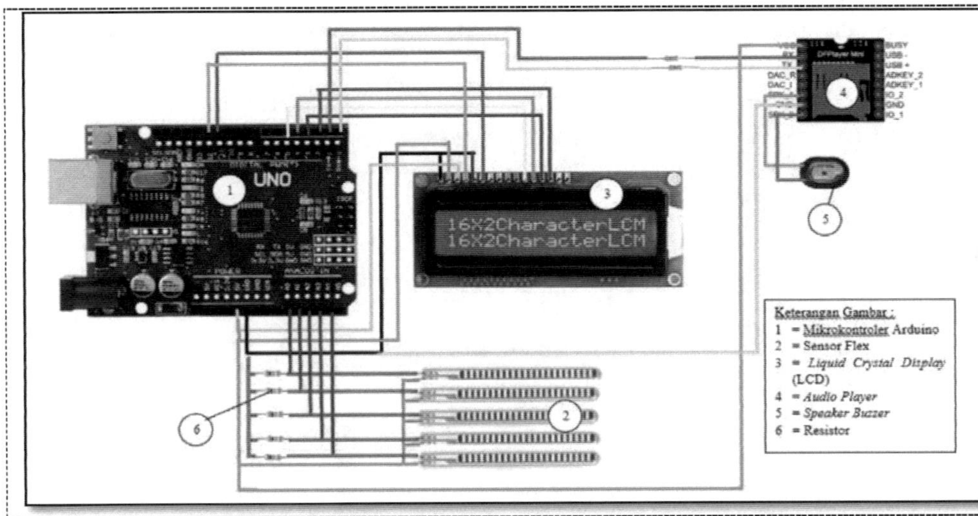

Figure 1. Hardware Design of the Sign Language Interpreter Tool.

4. Glove design

These translator gloves are designed using motorcycle gloves that are already on the market. Next the hardware shown in Figure 1. werepaired on the glove as shown in Figure 2. Flex sensor as placed on each finger, so that every movement of the fingers that occurs can be detected directly. In addition, other hardware such as the Arduino, LCD, Buzzer Speaker, and mini Dfplayer are placed on the wrist that has been coated by a 15 cm hand bracelet. Fig. 2 is a translator glove prior to repair.

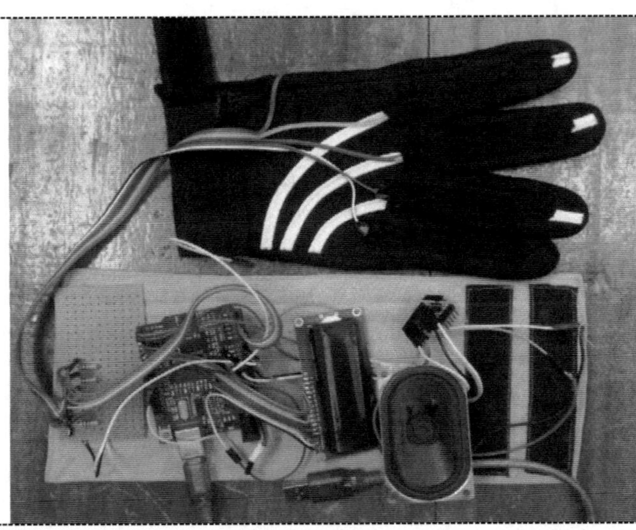

Figure 2. Glove Sign Language Translator.

IOP Conf. Series: Materials Science and Engineering **931** (2020) 012003 doi:10.1088/1757-899X/931/1/012003

5. Result and discussions

The data processing performed on the results and the discussion consists of two outcomes, namely the result of the design of the sign language interpreter and the method of Kansei engineering.

5.1. Declaration of the code

In the declaration of the code, previously identified the digital code displayed by the Arduino display based on finger movements made.Figure 3. Indicates the digital code generated by the flex sensor. The value is a digital code generated from the user's finger gestures. The code is the data entered in the main coding as shown in Figure 4.

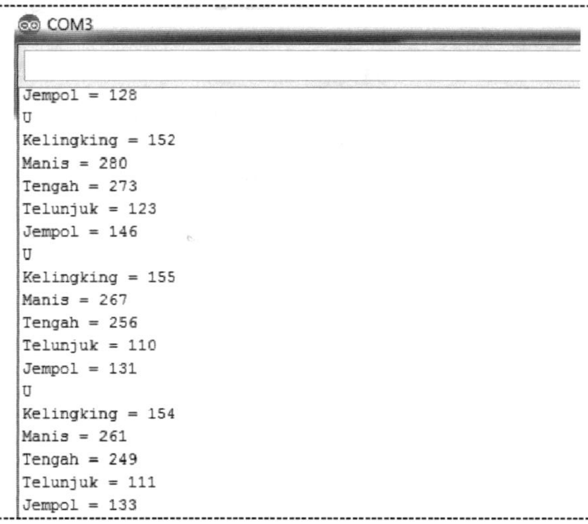

Figure 3. Digital code on Arduino view.

Figure 4. Primary coding on the Sign Language Translator.

Figure 4 is the main programming coding of sign language translators. The coding is formed in order to declare the order shown on the display by using the digital code value limit on the If program structure. If the digital code is within these limits, then the micro-controller will call the command and produce the output in the form of sound through speaker buzzer and display text on the Liquid Crystal Display (LCD).

5.2. Kansai engineering method

Table 2 shows recapitulation of results obtained from semantic differential questionnaires that have been filled by experts consisting of 3 persons and 3 teachers of Special School.

Table 2. Recap of Questionnaire Results Glove Sign Language Translator.

No	*Kansei Word*	Scale					*Kansei Word*
		1	**2**	**3**	**4**	**5**	
1	Cheap	0	2	4	0	0	Expensive
2	Simple	2	2	0	2	0	Complex
3	Light weight	1	3	1	1	0	Heavy
4	Practical	2	2	2	0	0	Difficult
5	Innovative	5	0	1	0	0	Plain
6	Rigid	0	3	3	0	0	Flexible
7	Futuristic	5	0	1	0	0	Conventional
8	Informative	0	3	2	1	0	Confusing
9	Natural	3	0	1	2	0	Engineered
10	Comfortable	2	3	1	0	0	Uncomfortable
11	Enthusiastic	2	4	0	0	0	Boring
12	*Colorful*	0	0	5	1	0	Plain color

Table 2. Shows the answers that respondents fill in based on the emotions they feel. The lower the scale chosen by the respondent, the Kansei word tends to correspond to the left side. However, the scale of the selected respondents will be in accordance with the Kansei word on the right side. While the value 0, 2, 1, 3, and other values contained in the column scale are the number of respondents who choose the scale.

5.3. Preference analysis based on kansai word categories

5.3.1. Cheap category

Sign language interpreters have prices ranging from RP 1,500,000.00 to RP 1,700,000.00. The bid price provides different preferences among the people, but the community assumes that the product has a cheap and normal price, as shown in Figure 5.

Figure 5. shows that as many as 67%, respondents tend to choose a 3rd scale stating that the price offered tends to be at a standard or normal price, neither cheap nor expensive. However, as many as 33% of respondents think that translator tool products have low prices. While the scale 4 and 5 indicate that the product has a high price and no respondent chooses the scale. The large number of respondents who say that the product is at a standard price, indicating that the price of the product does not need to be changed.

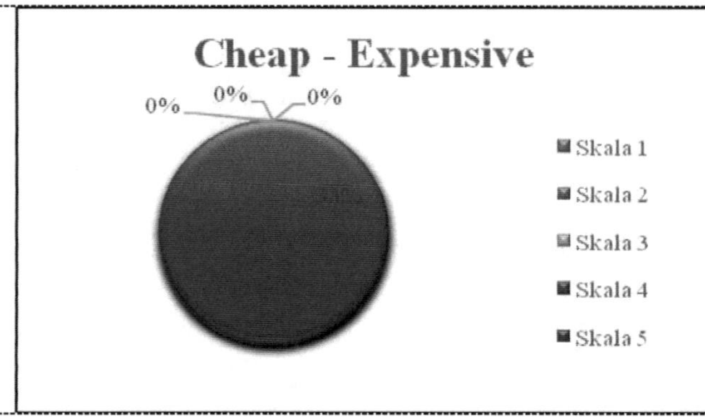

Figure 5. Kansei Word Trend For Cheap-Expensive (Price).

5.3.2. Simple-complex category

This category shows the shape of the sign language translator tool tend to be simple or complex when the respondent use it.

Figure 6 shows that 67% of respondents tend to say that the sign language interpreter has a simple form. The ease of using a sign language interpreter tool like using ordinary gloves makes the respondents choose scale 1 as much as 34% and scale 2 as much as 33%. However, as many as 33% of respondents feel disturbed by the cables that are on the upper side of the hand that makes the user difficult to move freely due to fear if the cable will break or electro-cut the user so that respondents choose a scale of 4 to express the emotion that is felt when using the tool.

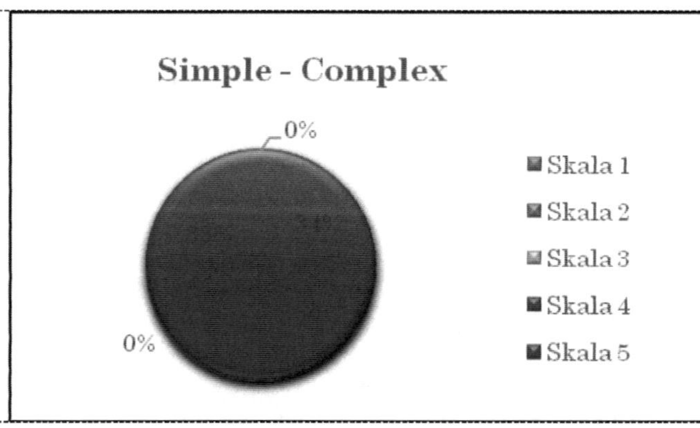

Figure 6. Kansei Word Trend For Simple-Complex.

5.3.3. Light-Heavy category

Figure 7 demonstrates the respondent's tendency on the light-weightedness of the sign language interpreter when used.

Figure 7 shows that respondents tend to choose a scale of 1% to 16% and a scale of 2 as much as 50% which leads to lightweight product traits. The lightness of the sign language interpreter tool affects the feelings of consumers who are not easily tired or stiff when using the tool of sign language interpreter in a long time. While 17% of respondents chose scale 3 which stated that the weight of the product is in normal position, not too heavy or not too light and 17% of respondents chose scale 4 which stated that the product is in heavy size. This indicates that the product of the sign language interpreter does not need improvement in this category.

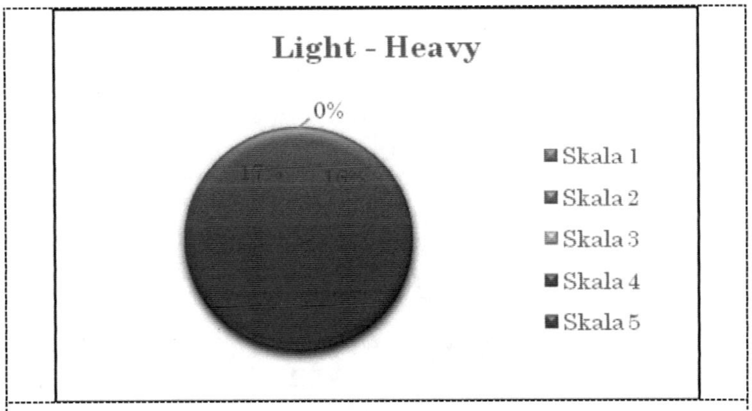

Figure 7. Kansei Word Trend For Light-Heavy.

5.3.4. Stiff-flexible category

This category tends to focus on the stiffness of gloves when users move their fingers to signal. This is shown in Figure 8.

Figure 8 shows that 50% of respondents chose scale 2 because they thought that the gloves were rigid, so the fingers were not free when moved and formed language signals so that repairs needed to be re-applied to the gloves. In addition, the movement of the finger is not free, influential with the finger gesture to be shown. This is evidenced by the inconsistency of the results displayed by the screen on Alphabets E and O. The screen sometimes displays the letter O when the respondent's practices the Alphon E signals. While the other 50% respondents chose a scale of 3 indicating that the product is normal. Scales 1, 4 and 5 are not an answer choice that suits the user's emotions when using a sign language translator product.

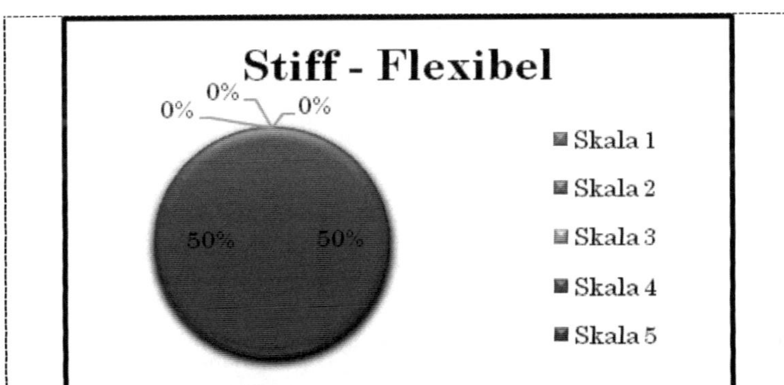

Figure 8. Kansei Word Trend For Stiff-Flexible.

5.3.5. Color category

Figure 9 shows the trend toward the color of the sign language translator tool that has been designed by the researcher.

Based on figure 9, 83% of respondents chose a scale of 3 that explains that the sign language interpreter has a normal color or is on a standard scale. However, 17% of respondents thought that the color of the glove is too monotonous, namely black colored so that the impression given on the glove tends to be casual when viewed in the colour side of the tool. Therefore, it is necessary to improve the gloves of translators with colors that are not monotonous.

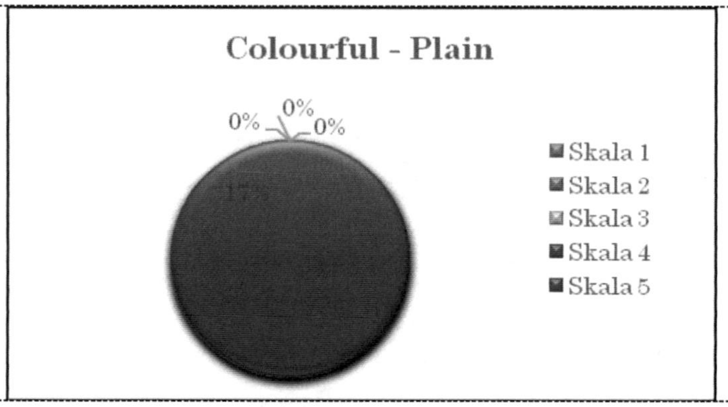

Figure 9. Kansei Word Trend For Colorful-Plain.

Based on the results of the discussion on the user, obtained some product evaluation results disclosed by the user when using the product Translator tool language that has the concept of data glove. Table 3 shows an evaluation of the Sign Language Tool.

Based on Table 3, it can be seen, there are several proposed improvements proposed by experts both on the tool form and function. Therefore, some basic improvements that have been donedo not change the aesthetics of the initial design of products that can disrupt the system work itself. Here are the improvements made and can be seen in Figure 10.

1. Stiff gloves were replaced with flexible and blue color's gloves. According to Cerrato [13], the blue color is able to provide calmness, relieving stress and fear. In addition, it can slow down the heart pressure and help to reduce appetite. Due to this reason, the researcher supports the respondent's suggestion to use blue-colored gloves.
2. Electronic components contained on the wrist closed with a small box.

Table 3. Product Evaluation of the Sign Language Interpreter Tool.

No	*Kansei Word*	Product Evaluation	Improvements to the Product
1	Cheap- expensive	Prices are convenient with the economic standards of society	-
2	Simple-complex	The shape of the product is appropriate and not complicated	-
3	Light- heavy	The product tends to be heavy due to the Mechatronic device attached to the user's arm.	-
4	Practice- difficult	The use of the product is very easy and appropriate.	-
5	Inovative-conventional	Products are innovative and new	-
6	Stiff – Flexible	Need to make changes to the gloves, the user is difficult to move the finger because the gloves are stiff.	Gloves have been replaced with gloves that are more flexible and adjust the user's hands
7	Futuristic – old fashion	The product is already futuristic because it is able to detect alphabetic shape well using sensor help.	-
8	Informative–confuse	The translator tool has not been able to detect hand movements and is only able to detect letters only	-

9	Natural- artificial	The product is natural	-
10	Comfortable-uncomfortable	Cable on the back of the palm should be closed because the user is afraid if the cable can make the user electrocuted	The wires on the surface have been covered by fabric.
11	Anthusiastic – boring	The product gives the user a passion when using it	-
12	*Colourful* - plain	Preferably the color of the gloves is adjusted more colored so that the product is able to provide an attraction for the user when viewing.	The color of the gloves has been replaced with a blue color adjusted by the user's wishes.

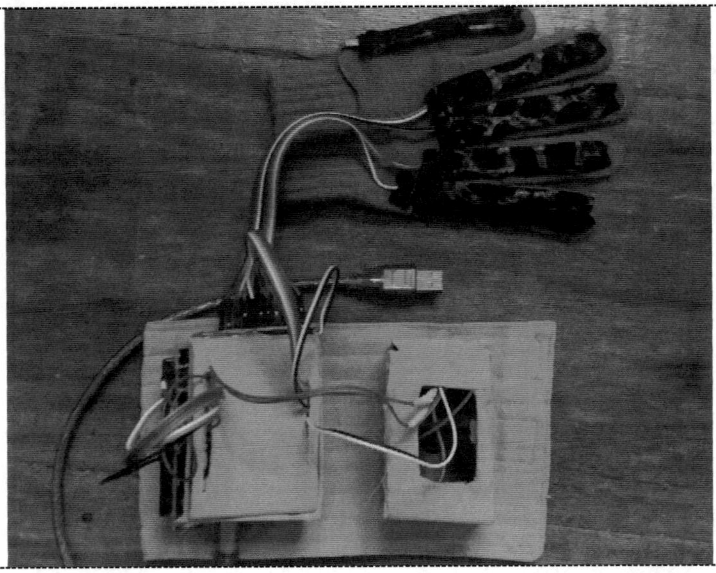

Figure 10. Improved Sign Language Improvement Tool.

Figure 10 represents an improvement that has been made on the glove sign language interpreter in accordance with the proposed improvement of the respondents. These improvements are made taking into consideration the product functions and needs of the users. In addition, these improvements are made on demand from consumers so that the product has been adapted to the consumer's emotions.

6. Conclusion

This research has been successfully done by designing a sign language interpreter tool tailored to the needs, wants, and emotion of users when using the product. The design of the tool using the glove data concept was done by identifying the value of digital code displayed by Arduino, so that the value can be a limitation in the formation of signal letter. In addition, the translation done by this tool only focuses on the vowels only, namely A, I, U, E, and O and has provided positive emotions from the main user of the signal interpreter tool. However, some users propose better improvements to the gloves to be flexible, light, and fashionable so that users are able to move their fingers freely. The suggestion is implemented by the researcher so that the sign language translator is really in accordance with the wishes of the customer.

IOP Conf. Series: Materials Science and Engineering **931** (2020) 012003 doi:10.1088/1757-899X/931/1/012003

7. References

[1] World Health Organization 2017 http://www.who.int/mediacentre/factsheets/fs300/en/ [Accessed 9 Maret 2017].

[2] El-Gayyar M M, Ibrahim A S and Wahed M E 2016 Translation from Arabic Speech to Arabic Sign Language Based on Cloud Computing *Egyptian Informatics Journal,* pp. **295-303**

[3] Abdallah E E and Fayyoumi E 2016 Assitive Technology for Deaf People Based on Android *Procedia Computer Science vol 94* **295-301**

[4] Bidin A S, Lokman A M, Rahim A and Tsuchiya T 2017 Initial Intervetion Study of Kansei Robotic Implementation for Elderly *International Journal Advanced Research in Computer and Communication Engineering vol 4* **430-434**

[5] Aziz A A, Moganan F F, Ismail A and Lokman A M 2015 Autistic Children's Kansei Response Towards Humanoid-Robot As teaching Mediator *Procedia Computer Science vol 76* **488-493**

[6] Shergian A and Immawan T 2015 Design Innovative Alarm Clock Made from Bamboo with Kansei Engineering Approach *Agriculture and Agricultural Science Procedia vol 3* **184-188**

[7] Nagamachi M 2011 Kansei/Affective Engineering *United States of America: Taylor and Francis Group LLC*

[8] Gunawan A A and Salim A 2012 Pembelajaran Bahasa Isyarat dengan Kinnect dan metode Dynamic Time Wrapping *Jurnal Matematika Statistik vol 3* **77-84**

[9] Rakhman J P, Ramadijanti N and Satriyanto E 2010 Translasi Bahasa Isyarat *Teknik Informatika-ITS*

[10] Kuroda T, Tabata T, Ikuta H and Mukarami M 2004 Consumer Price Data-glove for Sign Language Recognition *Virtual Reality and Association of Technology* **253-258**

[11] Razza B and Paschoarelli L C 2015 Affective Perception of Disposable Razors: A Kansei Engineering Approach *Procedia Manufacturing 3* **6228-6236**

[12] Wicaksono F 2015 Penerapan Kansei Engineering pada Rancangan Antarmuka E-learning Berbasis Web (Studi Kasus: STMIK CIC Cirebon). *Tesis Universitas Komputer Indonesia*

[13] Cerrato H 2012 The Meaning of Colors *Herman Cerrato Graphic Designer*

IOP Conf. Series: Materials Science and Engineering **931** (2020) 012004 doi:10.1088/1757-899X/931/1/012004

Design and development wire cutter CNC for Styrofoam product

Syahriza, U Aulia, L Muspika
Department of Mechanical and Industrial Engineering, Universitas Syiah Kuala, Darussalam, Banda Aceh

E-mail: syahriza@unsyiah.ac.id

Abstract. Cutting Styrofoam manually will result in inaccurate cutting and dissimilar shape. To upgrade the quality of the product, a Styrofoam cutting machine which can cut automatically is needed. Thus, the aim of this research is to create a Styrofoam cutting machine using CNC 2 Axis Wire Cutter which can cut the material according to the CAD drawing module, and accelerate the cutting process. The control system of the machine uses the Arduino UNO microcontroller. The cutting process is performed by sending G-Code file to the microcontroller through the Universal G-Code Sender software. Afterwards, the microcontroller sends a signal to activate the stepper motor, and it is generated actuator movement according to the drawing in G-code file. The optimal Styrofoam cutting resulted by using CNC Wire Cutter with 0.2 mm nickel wire, 220oC, and 2 cm thickness.

1. Introduction

The development of the industry is growing continuously. As an example, making decoration from Styrofoam Decoration can be used in various things, placards, art work, making safety goods on shipping, and others. The use of CNC Wire Cutter in making decoration will shorten the time needed [1]. Currently many Styrofoam craftsmen in Indonesia use manual Styrofoam cutting tools, namely cutters and hot wire [2]. A good and planned manual cutting of Styrofoam involves a number of other activities as supporting activities such as measurements, sketches and others. To overcome the large number of product ordered in different shapes, it will be a problem as it requires concentration in sketches, measurements and cuts [3].

To solve this problem, CNC Wire Cutter is used as a substitute for cutting tools made of cutter blades, multiplex saws or similar tools which are inappropriate in terms of cutting results compared to CNC Wire Cutter tools.

The working principle of the Cutter CNC 2 axis Styrofoam cutter is by converting electrical energy into heat energy. This CNC Wire Cutter already uses numerical control using CNC programming, and the cutting precision is much faster. Wire Cutter CNC machine can cut as designed by CAD drawing modules that work automatically with G-code input from the computer.

2. Literature review

CNC stands for (Computer Numerically Controlled,) beginning in 1952. In general, CNC machine tools work similarly to the conventional machine tools. The function of CNC replaces the human work in conventional machine tools such as the work of setting tools or adjusting the chisel movement to the

IOP Conf. Series: Materials Science and Engineering **931** (2020) 012004 doi:10.1088/1757-899X/931/1/012004

position ready to cut, the cutting motion and movement back to the initial position, and others. CNC operating system parameters can be changed through the appropriate software program [4].

2.1. CNC machines

There are several CNC machine that has already existed in the world (Figure 1). The 5 axis CNC machine works with 5 limbs, which are X, Y, Z axis for linear motion then axis A for horizontal rotating motion, and C axis works for vertical rotating motion. CNC Lathe has two-way movement (2 axis) with numerical control (CNC). CNC Plasma Cutting Machine The construction works on a CNC plasma plate cutting machine, and very similar to a 2.5 axis CNC engraving machine or a 3-dimensional scanner. The only difference is that this plasma machine uses gas (plasma) throwers, not chisels.

Figure 1. CNC milling machine

2.2. Hot-wire cutting (manual)

This model adopts a layered object manufacturing technique (LOM) used for fast manufacturing. It is to build a Styrofoam model that can be used (about 30-100 mm) in a Styrofoam layer cut with a hot wire and brought closer to form the intended object, an electrically heated wire (Figure 2).

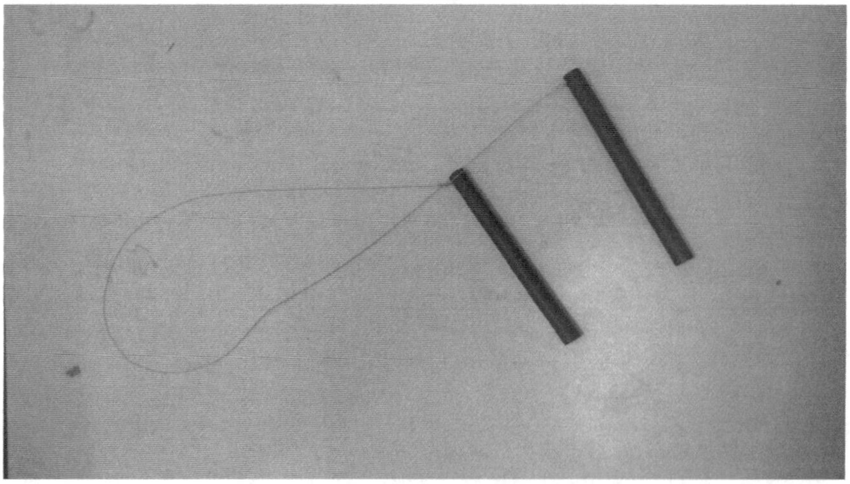

Figure 2. Hot-wire cutting (manual)

In order to cut it more accurately, it requires a lot of skills and manpower to achieve the desired results. This research is an effort to build a system that serves the needs we want which is to be faster and more accurate [5].

3. Research methods

The flow of the research process that will be carried out can be seen in the following flowchart Figure 3.

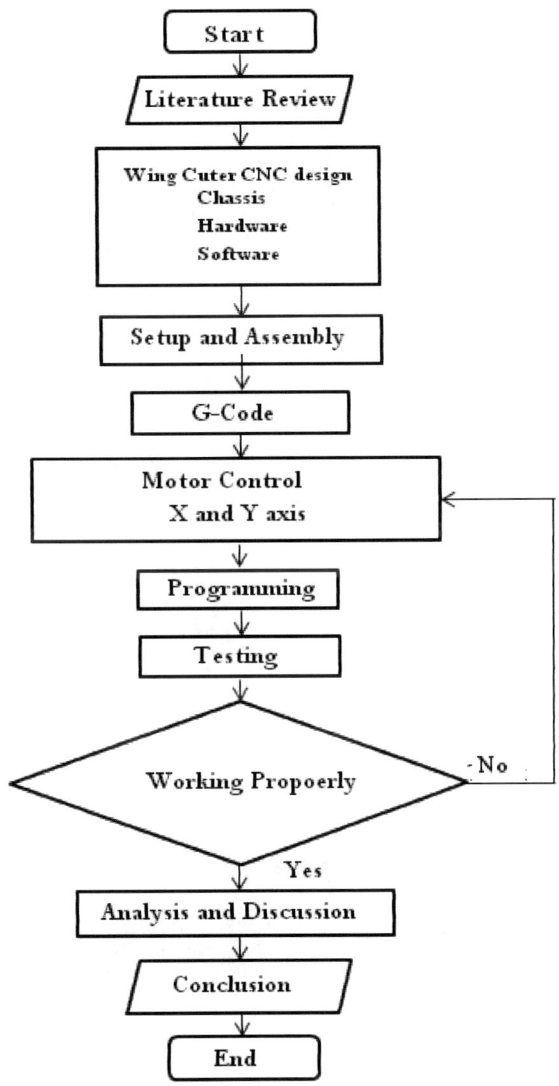

Figure 3. Research flowchart

3.1. Stages of research implementation

The results of the research are as follows:

a. Design and manufacture of chassis in CNC Wire Cutter (Figure 4).
b. Set up on a Wire Cutter NC using the program MakerCam, Arduino, G-Code and GRBL (Figure 5).

IOP Conf. Series: Materials Science and Engineering **931** (2020) 012004 doi:10.1088/1757-899X/931/1/012004

c. Test on CNC Wire Cutter.
d. Analyse and discuss research data, and make conclusions

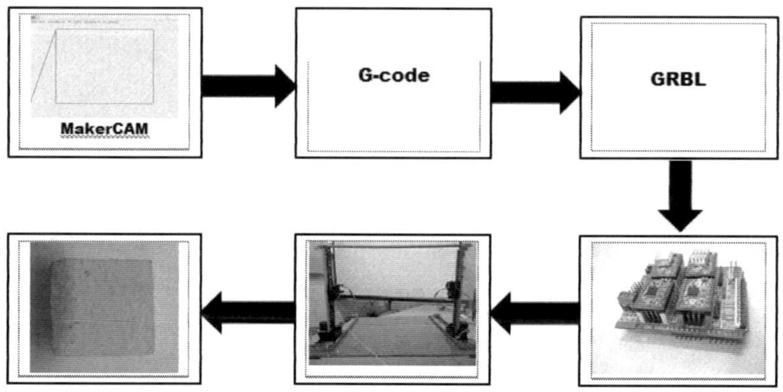

Figure 4. Block Design Process Diagram

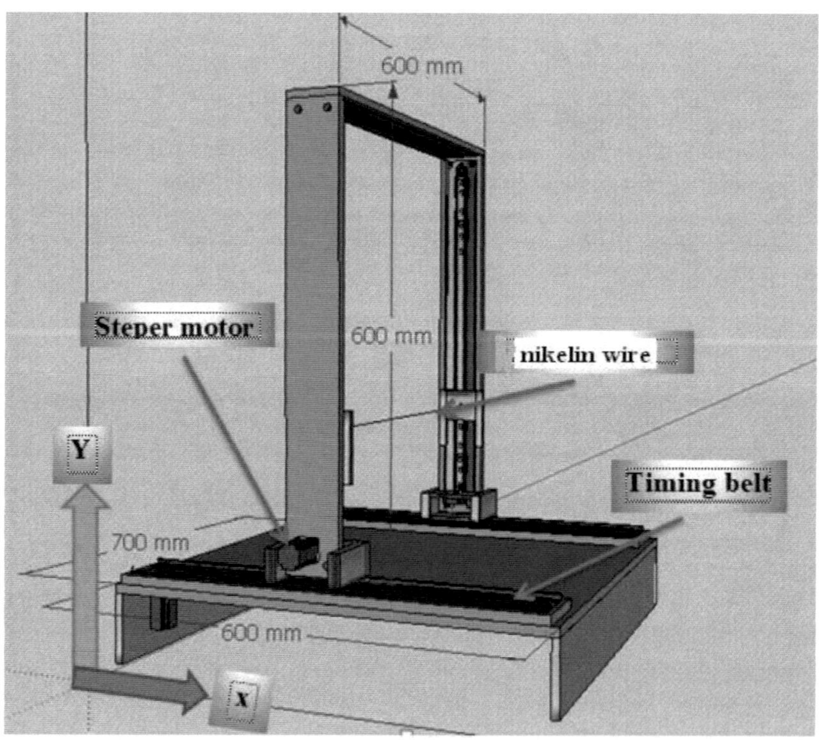

Figure 5. Wire CNC Cutter design

3.2. CNC wire cutting machine
The equipment used in CNC Wire Cutter research is as follows

a. Nickel wire
Nickel wire is used to heat elements in styles using Styrofoam. There are two kinds of nickel wire and nickel flattened / flattened. Nickel wire is used to distribute heat and has a melting

point of 1400 ° C. The normal to cut Styrofoam is 200° C - 300° C. There are several sizes of nickel wire; 0.2 mm, 0.3 mm, 0.5 mm, 0.6 mm.

b. Stepper motor
Stepper motor is an electromechanical device that cannot move on their own and must be connected to the Arduino Uno and CNC Shield. Their movements are per step according to their specifications, and moving from one step to the next requires time and produces large torque at low speeds. The stepper motor used in this study has 200 steps per motor rotation.

Figure 6. Bipolar stepper

c. CNC shield
CNC Shield is an intuition between Arduino UNO and stepper motors with the ability to move 4 stepper motors to drive 2 Axis (Figure 7).

Figure 7. CNC shield

d. Arduino UNO
Arduino UNO as shown in Figure 8 has 14 digital inputs / outputs (6 can be used as PWM outputs), 6 analogue inputs, a 16MHz Crystal oscillator, a USB connection, a power jack, a label, and a reset button.

Figure 8. Arduino UNO

e. Testing cutting process
Tests to be carried out with the following sizes (Table 1).

Table 1. Accuracy Testing

No	Dimension	h (mm)	T (°C)	V (mm / minute)	Error
1	Square	20	310	250	5.2
2	Square	20	300	250	4.4
3	Square	20	290	250	4
4	Square	20	280	250	3.6
5	Square	20	270	250	3.1
6	Square	20	260	250	2.6
7	Square	20	250	250	2
8	Square	20	240	250	1.4
9	Square	20	230	250	1
10	Square	20	220	250	0.5

T: Temperature (° C); h: thickness (cm); V: speed (cm / s);
E: error (mm)

f. Mechanism of moving CNC wire cutter
This test was carried out in order to find the right temperature to cut 2 cm thick Styrofoam. Styrofoam was cut using a timing belt to move the Axis driven by 4 stepper motors with 2 movements, X, Y, the movement had been programmed using the software MakerCAM, Arduino UNO, Xloader and GRBL.

g. Testing of the same shape
The optimal result of cutting rectangular shape of 2 cm thick Styrofoam was with a temperature of 220oC, cutting speed of 250 mm / min and cutting edge of 0.2 mm. Cutting results of 2 cm thick Styrofoam produce the smallest error value of 0.5 mm and the biggest error of 5.2 mm with a temperature of 310°C, and cutting speed of 250 mm / min.

h. Testing with different forms
This test was carried out to determine the ability of CNC Wire Cutter with different shapes, rectangles, hexagons, triangles and circles.

IOP Conf. Series: Materials Science and Engineering **931** (2020) 012004 doi:10.1088/1757-899X/931/1/012004

i. Square testing
Cutting Styrofoam using G-code (Figure 9). The G-code file was programmed into GRBL and visual results as seen in Fig. 9. In the visual image, a red dot shows the zero point of the G-code image, the blue track will move according to the path and the red line in the G- image X and Y is axis line code. After the visualization results are in accordance with the design in the MakerCAM software, the G-code file is sent to the microcontroller and the cutting process runs automatically.

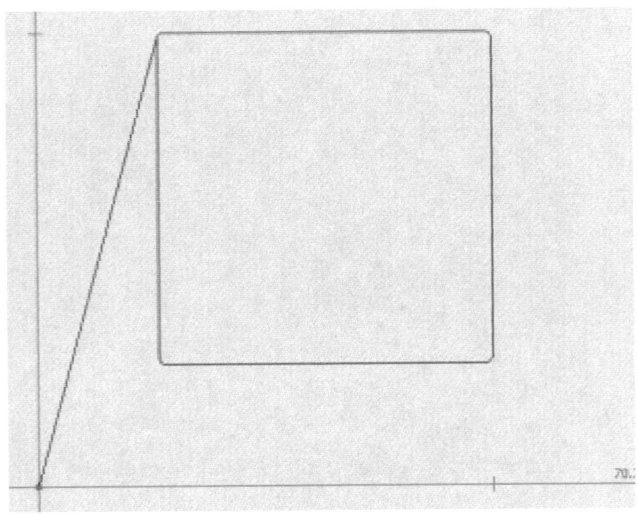

Figure 9. G-code file

In this test, the 2 cm thick Styrofoam was cut with a temperature of 250oC, cutting speed of 250 mm / min and cutting edge of 0.2 mm. Cutting results show an error value of 0.5 mm and can be seen in Figure 10.

Figure 10. Square

IOP Conf. Series: Materials Science and Engineering **931** (2020) 012004 doi:10.1088/1757-899X/931/1/012004

j. Hexagonal testing
In this test, the 2 cm thick Styrofoam was cut with a temperature of 250oC, cutting speed of 250 mm / min and cutting edge of 0.2 mm. Cutting results show an error value of 0.5 mm and can be seen in Figure 11.

Figure 11. Hexagonal

k. Triangular testing
In this test, the 2 cm thick Styrofoam was cut with a temperature of 250°C, cutting speed of 250 mm / min and cutting edge of 0.2 mm. Cutting results show an error value of 0.5 mm and can be seen in Figure 12.

Figure 12. Triangular

IOP Conf. Series: Materials Science and Engineering **931** (2020) 012004 doi:10.1088/1757-899X/931/1/012004

4. Conclusion

Based on the results of this research, the following conclusions can be drawn:

1. The author has succeeded in designing and constructing a 2 Axis CNC Wire Cutter, Styrofoam cutting converts electric current into heat using nickel wire.
2. Cutting CNC Wire Cutter using software MakerCAM, G-code, Arduino UNO, GRBL works well.
3. CAD systems can use line, circle, arc entities to produce product geometry.
4. The optimal result cutting the same shape Styrofoam of 2 cm thick is using the temperature of 220oC, and a speed of 250 mm / min. The biggest error is 5.2 mm with temperature of 310°C.

5. Reference

[1] Sulkowski W W Nowak K Sulkowska A Wolinska A Bajdur W M Pentak D Mikula B 2009 Study of the Sulfonation of Expanded Polystyrene Waste and of Properties of the Products Obtained *Pure Appl.Chem* **81** 2417-2424

[2] Stege 2011 Method of Producing Test Components by A Hot Wire Cutter *Canada Patent CA2744470 A1* 29 December 2011

[3] Wijaya P D Rivai M Tasripan 2017 Rancang Bangun Mesin Pemotong Styrofoam 3 Axis Menggunakan Hot Cutting Pen dengan Kontrol PID **6**

[4] Information on http://meirasta.blogspot.co.id/2015/03/pengertian-mesin-cnc.html

[5] Cheng S N Hot-wire Cutter United States Patent US 8278585 B2 0310 2012.

[6] Aitchison D R Bain J Brooks H Taylor M 2007 Transient Temperature Effects in Hot-wire Plastic Foam Cutting *4th Int. conf. on Design and Production of Machines & Dies/Molds* 219-224

[7] information on www.dany.setiawan@windowslive.com.

IOP Conf. Series: Materials Science and Engineering **931** (2020) 012005 doi:10.1088/1757-899X/931/1/012005

Higher Resolution Input Image of Convolutional Neural Network of Reinforced Concrete Earthquake-Generated Crack Classification and Localization

Muammar Sadrawi[1], Husaini[1], Jalaluddin Yunus[1], Irwansyah[1], Maysam F. Abbod[2] and Jiann-Shing Shieh[3]

[1] Department of Mechanical and Industri Engineering, Faculty of Engineering, Universitas Syiah Kuala, Banda Aceh, Indonesia

[2] College of Engineering, Design and Physical Sciences, Brunel University London, Uxbridge UB8 3PH, UK

[3] Department of Mechanical Engineering, Yuan Ze University, Chung-Li Dist., Taoyuan City, 32003, Taiwan

E-mail: husainiftm@unsyiah.ac.id

Abstract. According United States Geological Survey, Aceh is the northwestern part in Indonesia that has been affected by numerous strong earthquakes since 2004 tsunami. These earthquakes have generated massive impact to the buildings around the area, especially for the reinforced concrete based buildings. One of the most important problems to the reinforced concrete is the earthquake-generated crack. In this study, the dataset from the normal and cracked reinforce concrete are collected by taking the normal and cracked images. Several convolutional neural network models are implemented such as LeNet based models. These models are initially applied to recognize either normal or cracked conditions. Eventually, for the last stage, the localization of the crack is visualized by imposing the original images. For the localization, this study also evaluates the relatively smaller and bigger cracks. The results show the higher input image with modified LeNet generates better results compared to the basic model in superimposing the localized crack.

1. Introduction

Aceh was hit by tsunami generated by a 9.1 Mw earthquake on 2004. According United States Geological Survey (USGS), many relatively big earthquakes struck the area afterwards. In more detail, several earthquakes that have magnitude higher than 6.5 Mw occurred after the largest one. It can be seen on Table 1 based on the region shown in Fig. 1 [1].

Crack is one of the biggest problems caused by the earthquake. Several studies have investigated crack effects specifically to building related applications. The utilization of Gabor filter to detect crack in the tunnel has been performed in a study [2]. Also, some learning methods have been implemented for the concrete crack detection [3]. Furthermore, signal processing-based algorithm was also used for the crack detection in the building [4].

Crack may occur in other applications apart from the building. Crack can also occur in aerospace material [5]. In this study thermoelastic stress analysis and acoustic emission were involved for the fatigue crack initiation and propagation evaluation. Digital image correlation was performed in order to solve the crack problem in asphalt [6]. Crack has been studied also in polymer [7] and aluminum alloy [8].

Artificial intelligence (AI) has been widely used in applications especially for image processing [9] and unbalanced and ensemble modeling [10]. For biomedical engineering, AI has been used for biomedical image processing-based application [11] and signal processing-based application such as anesthesia [12] and arrhythmia [13, 14].

This study is a two-stage evaluation. The first stage, the classification of the cracked and normal reinforced concrete is evaluated. Meanwhile, the next stage is to investigate the location of the crack when the cracked condition appears on the previous stage. The evaluation proposes several modified CNN models with higher resolution input image in superimposing the original image.

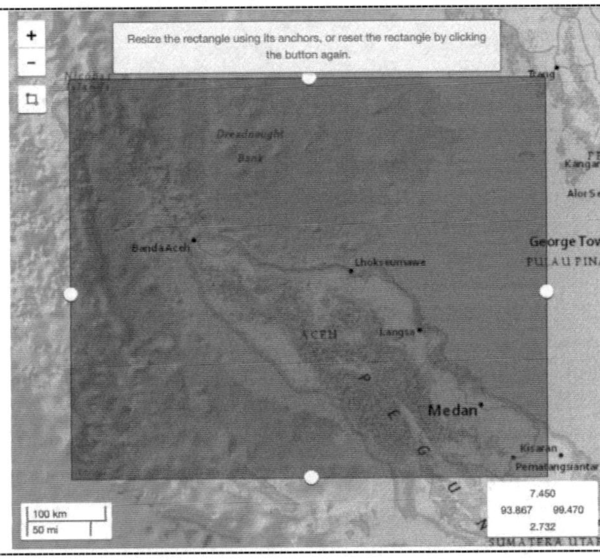

Figure 1. USGS evaluated earthquake region around Aceh [1].

Table 1. Largest earthquakes occurred in area of Aceh [1].

Time	Latitude	Longitude	Depth	Magnitude	Magnitude Type
2004-12-26T00:58:53.450Z	3.295	95.982	30	9.1	mw
2008-02-20T08:08:30.520Z	2.768	95.964	26	7.4	mwc
2010-05-09T05:59:41.620Z	3.748	96.018	38	7.2	mwc
2005-02-26T12:56:52.620Z	2.908	95.592	36	6.8	mwb
2011-09-05T17:55:11.220Z	2.965	97.893	91	6.7	mww
2015-11-08T16:47:02.160Z	6.8431	94.648	10	6.6	mww
2016-12-06T22:03:33.390Z	5.2834	96.1678	13	6.5	mww
2008-03-29T17:30:50.150Z	2.855	95.296	20	6.3	mwc
2005-03-30T16:19:41.100Z	2.993	95.414	22	6.3	mwb

2. Dataset and methodology

The dataset uses in this study is achieved by collecting the reinforced concrete images. It contains the normal and cracked reinforced concrete. The raw figure is taken by iPad Air (Apple Inc.). Total images for this study is 573 images with the raw size of 1936 pixel × 2592 pixel. The images utilized for the training is 375 images. Meanwhile, a set of 198 images for normal and cracked concrete is used for the testing the CNN models. Other setting is set as default. This study uses MacBook Pro 2.6 GHz Intel Core i7. Furthermore, Python version 3.6.8, OpenCV version 3.4.2 and Keras 2.1.2 with Tensor Flow 1.4.1 [15] are the utilized software.

Table 2. Confusion matrix evaluation

		Predicted	
		Cracked	Normal
Reference	Cracked	True Positive (TP)	False Negative (FN)
	Normal	False Positive (FP)	True Negative (TN)

This study evaluation utilizes the confusion matrix, as shown on Table 2, with sensitivity and specificity. In more detail, the sensitivity is the ratio of the cracked material, as the reference, is well predicted to the cracked material, true positive (TP), to the cracked material, also as the reference, is miss predicted to the normal material (FN), shown on Eq. 1. Meanwhile, the specificity is the ratio of the normal condition as the reference is predicted as the normal as well, true negative (TN) compared to the miss classified the reference of the normal condition as the cracked condition, that can be seen on Eq. 2.

$$Sensitivity = \frac{TP}{TP + FN} \tag{1}$$

$$Specificity = \frac{TN}{TN + FP} \tag{2}$$

```
Layer (type)                 Output Shape
=====================================================
conv2d_1 (Conv2D)            (None, 512, 512, 32)
_____________________________________________________
activation_1 (Activation)    (None, 512, 512, 32)
_____________________________________________________
max_pooling2d_1 (MaxPooling2 (None, 256, 256, 32)
_____________________________________________________
conv2d_2 (Conv2D)            (None, 85, 85, 64)
_____________________________________________________
activation_2 (Activation)    (None, 85, 85, 64)
_____________________________________________________
max_pooling2d_2 (MaxPooling2 (None, 42, 42, 64)
_____________________________________________________
conv2d_3 (Conv2D)            (None, 14, 14, 128)
_____________________________________________________
activation_3 (Activation)    (None, 14, 14, 128)
_____________________________________________________
max_pooling2d_3 (MaxPooling2 (None, 7, 7, 128)
_____________________________________________________
flatten_1 (Flatten)          (None, 6272)
_____________________________________________________
dense_1 (Dense)              (None, 256)
_____________________________________________________
activation_4 (Activation)    (None, 256)
_____________________________________________________
dense_2 (Dense)              (None, 2)
_____________________________________________________
activation_5 (Activation)    (None, 2)
```

Figure 2. Higher resolution input image LeNet model summary.

For the localization, the first activation or feature map is uniformly taken from every model. This condition can be also called as the heat map. This resized matrix will be reformed to the original image size. This procedure highly likely affects the localization marking for the crack.

3. Result and discussion

This study uses several concrete images from several buildings. The utilized models are the LeNet based models. The uniqueness of the LeNet model is one of the simplest architectures in deep learning algorithms especially for the CNN, which has 1,212,578 of trainable parameters. The activated convolution layers are mostly followed by the max pooling procedure. Moreover, the stride shape is one by one. This small number of strides will keep the image size to have sufficient size until the flattening layer. In this study, the LeNet model provides relatively good results in both training and testing, as it can be seen on Tables 4 and 5. However, for this LeNet model, the input size is only 150 × 150 pixels. In order to localize the crack from the original image, that has much bigger image size, 1936 × 2592 pixels, the first feature map generated by this model is only 148 × 148 pixels. The reduced size is due to the valid padding set to the model. This feature map will have problem in superimposing the original image. The detail about LeNet models can be seen on Figs. 2 and 3.

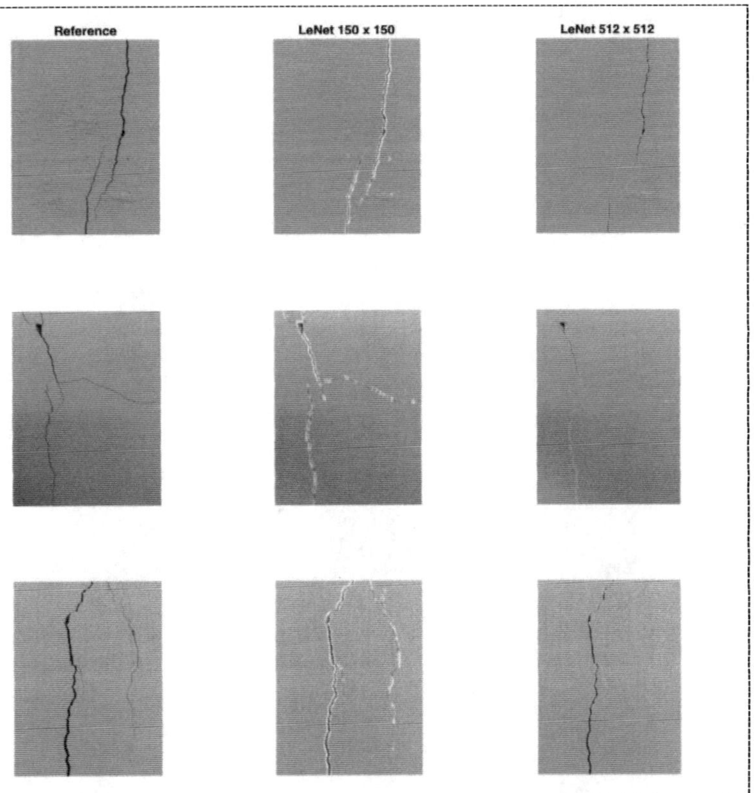

Figure 3. Smaller crack localization results of LeNet models.

Table 3. Training evaluation. TP = True Positive, FP = False Positive, TN = True Negative, FN = False Negative, SEN = Sensitivity, SPE = Specificity.

Input size	Confusion Matrix				SEN	SPE
	TP	FP	TN	FN		
150 x 150	122	0	238	13	0.9037	1.0000
512 x 512	128	2	236	7	0.9481	0.9916

Table 4. Testing evaluation. SEN=Sensitivity, SPE=Specificity

Input size	Confusion Matrix				SEN	SPE
	TP	FP	TN	FN		
150 x 150	102	0	75	19	0.8430	1.0000
512 x 512	121	0	75	0	1.0000	1.0000

In order to have better superimposing to the raw image, this study proposes a higher resolution input image. This proposed method has 512 × 512 pixels for the first feature map. The average value of the all channels will be the final feature map for superimposing the original image. This model has bigger architect, which is 1,699,650 trainable parameters. This modified algorithm even tough has slightly worse in specificity than the basic model, this updated model provides better sensitivity in training. Moreover, in testing, the higher resolution input image has an improved result for the sensitivity. The detail can be seen on Tables 3 and 4.

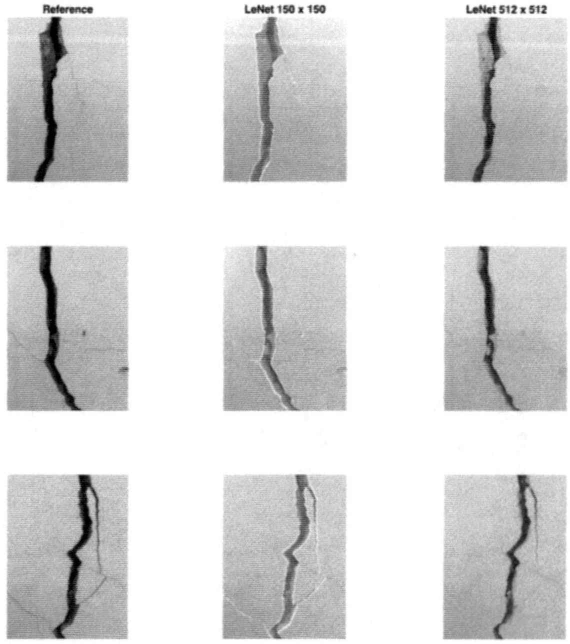

Figure 4. Bigger crack localization results of LeNet models.

For the crack localization it can be seen on Figures 4 and 5. Figure 4 provides the localization results for relatively smaller cracks. It can be seen that the higher resolution LeNet that has 512 × 512 pixels, produces much better result in superimposing the raw images compared to the basic LeNet model that has 150 × 150 pixels. The result of the bigger crack localization can be seen on Fig. 5. From this figure the higher resolution input image model produces a better localization compared to the basic model. Especially for the bigger crack cases, the model is still able to differ the surface damage and the obvious crack as shown by the heat map distribution in the superimposed images.

4. Conclusion

This study evaluates the crack classification and localization. It initially classifies the material either normal or cracked concrete. Furthermore, if the condition is detected as the crack material, the crack localization will be performed sequentially. For the results, it can be seen that the higher resolution input image model provides a better classification and localization results. This study with some further works can be applied for the AI based automatic crack detector.

Acknowledgement

The authors would like to express their gratitude to Universitas Syiah Kuala for financial support partly for research through the Professor's research grant No. 03 / UN11.2 / PP / PNBP / SP3 / 2019.

References

[1]https://earthquake.usgs.gov/fdsnws/event/1/query.csv?starttime=2004-07-12%2000:00:00&endtime=2019-07-19%2023:59:59&maxlatitude=7.45&minlatitude=2.732&maxlongitude=99.47&minlongitude=93.867&minmagnitude=5&orderby=magnitude (Accessed on July 19th 2019).

[2] Medina R, Llamas J, Gómez-García-Bermejo J, Zalama E, Segarra M. *Crack detection in concrete tunnels using a gabor filter invariant to rotation*. Sensors. 2017;17(7):1670.

[3] Wang B, Zhao W, Gao P, Zhang Y, Wang Z. *Crack Damage Detection Method via Multiple Visual Features and Efficient Multi-Task Learning Model*. Sensors. 2018 Jun;18(6):1796.

[4] Mu W, Sun J, Liu G, Wang S. *High-Resolution Crack Localization Approach Based on Diffraction Wave*. Sensors. 2019 Jan;19(8):1951.

[5] Middleton CA, McCrory JP, Greene RJ, Holford K, Patterson EA. *Detecting and Monitoring Cracks in Aerospace Materials Using Post-Processing of TSA and AE Data*. Metals. 2019 Jul;9(7):748.

[6] Hasheminejad N, Vuye C, Margaritis A, Ribbens B, Jacobs G, Blom J, Dirckx J, Vanlanduit S. *Investigation of Crack Propagation and Healing of Asphalt Concrete Using Digital Image Correlation*. Applied Sciences. 2019 Jan;9(12):2459.

[7] Husaini, Kishimoto K, Notomi M, Shibuya T. *Fracture behaviour of PC/ABS resin under mixed-mode loading*. Fatigue & Fracture of Engineering Materials & Structures. 2001; 24(12): 895-903.

[8] Husaini, Kishimoto K, Hanji M, Notomi M. *Investigations of The Mixed Mode Crack Growth Behavior of An Aluminum Alloy*. Journal of Engineering and Applied Sciences. 2016;11(2): 885-890.

[9] LeCun Y, Bottou L, Bengio Y, Haffner P. *Gradient-based learning applied to document recognition*. Proceedings of the IEEE. 1998 Nov 11;86(11):2278-324.

[10] Sadrawi M, Sun WZ, Ma M, Yeh YT, Abbod M, Shieh JS. *Ensemble Genetic Fuzzy Neuro Model Applied for the Emergency Medical Service via Unbalanced Data Evaluation*. Symmetry. 2018;10(3):71.

[11] Sadrawi M, Fan SZ, Abbod MF, Jen KK, Shieh JS. *Computational depth of anesthesia via multiple vital signs based on artificial neural networks*. BioMed research international. 2015;2015.

[12] Ronneberger O, Fischer P, Brox T. U-net: *Convolutional networks for biomedical image segmentation*. International Conference on Medical image computing and computer-assisted intervention 2015 Oct 5 (pp. 234-241). Springer, Cham.

[13] Sadrawi M, Lin CH, Lin YT, Hsieh Y, Kuo CC, Chien J, Haraikawa K, Abbod M, Shieh JS. *Arrhythmia evaluation in wearable ECG devices*. Sensors. 2017;17(11):2445.

[14] Rajpurkar P, Hannun AY, Haghpanahi M, Bourn C, Ng AY. *Cardiologist-level arrhythmia detection with convolutional neural networks*. arXiv preprint arXiv:1707.01836. 2017 Jul 6.

[15] Abadi M, Agarwal A, Barham P, Brevdo E, Chen Z, Citro C, Corrado GS, Davis A, Dean J, Devin M, Ghemawat S. Tensorflow: *Large-scale machine learning on heterogeneous distributed systems*. arXiv preprint arXiv:1603.04467. 2016 Mar 14.

IOP Conf. Series: Materials Science and Engineering **931** (2020) 012006 doi:10.1088/1757-899X/931/1/012006

Study of suspended handle materials for reducing hand-arm vibration in backpack grass trimmer

A Z Mubarak, A Razali, M Rizal, K R Hambali
Department of Mechanical and Industrial Engineering, Universitas Syiah Kuala, Darussalam, Banda Aceh

E-mail: amir_zm@unsyiah.ac.id

Abstract. Backpack-type petrol-driven grass trimmers are a widely used grass trimmer. However, the high vibration on its handle causes machine operator to be at risk of "Hand Arm Vibration (HAV)" or vibration exposure on the arms that can damage the nervous system. This research aims to study the effect of suspended handle materials in reducing the vibration on the handle of the backpack-type grass trimmer. There are three types of materials used in this study, they are natural rubber, rubber silicone and foam EVA (Ethylene-vinyl acetate). The material is layered around the handle and then the vibration on the handle is measured. The result shows that foam EVA has better performance in reducing the vibration acceleration rather than natural rubber and rubber silicone.

1. Introduction

The backpack grass trimmer is most commonly used rather than other mowers. The high vibration in the handle of the machine exposed at the operator's hands poses a risk of Hand Arm Vibration (HAV) or vibration that can paralyze the nerves in the hand.

Previously there have been pieces of research on the effect of this grass trimmer on the operator. The grass trimmer operator in the area of Universitas Syiah Kuala has an activity limit of less than 1 hour per day based on vibrations received and it takes 30 minutes of rest time for each activity [1].

Several pieces of research have been conducted to overcome the magnitude of the vibration of this cutting machine to reduce the risk of usage in a long time. One of them has been studied making several types of handles with several different shapes and also different materials. The results of this study obtained a decrease in vibration acceleration of 76% of the engine tray handle, where the vibration acceleration on the default handle of 11.30 m/s2 can be reduced to 2.69 m/s2 with handles made [2].

Then there is a research with simulation method, where the simulated handle shape is covered with rubber material and PVC 0.007 Plasticized. The result of this study is that the effect of rubber material and PVC 0.007 plasticized is not too significant, even though the rubber material is more appropriate to be used for acceleration absorption compared to PVC material 0.007 plasticized [3].

There is also a balancing process on the handle of this type of grass trimmer with structural optimization. Through the analysis of vibration handles and pipes, it turns out to have a high effect on vibration. Then try to reduce the resonance frequency and response rate at high frequency to increase the vibration characteristics of the pipe handle given that it does not add to the total weight [4].

Vibration is a regular movement of objects or media in the direction back and forth from the position of balance. Vibration is a physical factor caused by subjects with oscillating vibrations, for example, machines, equipment or work tools that vibrate and expose workers through transmission [1]. These movements can be continuous and repetitive and can be irregular or random movements. Generally, vibrations are caused by the force that varies with time. Whereas what is meant by mechanical vibration is the vibration caused by the means and equipment of human activities. In occupational health, in general, is divided into two namely: the vibration of the whole body and the vibration of the arms of the arm.

Vibration in the whole body is a vibration that affects the whole body, which is delivered through the body parts of the labour that supports the entire body. For example: from the legs when standing, from the back when leaning, from the buttocks when sitting, or from the arms when leaning. Usually, the frequency of these vibrations is 5-20 Hz. Hand-arm vibrations are vibrations that travel through the hands due to the use of vibrating equipment. The frequency is between 20-500 Hz. The most dangerous frequency is at 128 Hz because the human body is very sensitive to this frequency. These vibrations occur in jobs such as chain saw, grass trimmers, grinding and drilling operators [3].

There are many different ways to control vibration. One way is to just absorb it. This is a process known as vibration damping. Over the years, many different materials have been developed that can help absorb and dampen vibrations from various sources. Dampening material is used to reduce or eliminate noise and vibration that occur in the industrial world caused by resonance. The dampening agent works to change the natural vibration frequency of the vibrating surface. The purpose of this research is to study the effect of suspended handle materials in reducing the vibration on the handle of the backpack-type grass trimmer.

2. Methods

The research is conducted by experimental method. The vibration is measured by using an accelerometer sensor (see Fig. 1) mounted on the handle of the grass trimmer. The result is collected by the data acquisition to processing the signal received by the sensor then the signal is converted to digital numeric that can be manipulated by a computer. The accelerometer can be used to measure acceleration, detect and measure vibrations, and measure the acceleration due to graphitization (inclination). An accelerometer can be used to measure vibrations in a car, engine and building. An accelerometer can also be applied to earthquake activity and electronic equipment. For more advanced applications, this sensor is widely used for applications [5].

Figure 1. Accelerometer

By using the Data Acquisition System (DAQ), as can be seen in Fig. 2, the vibration signals are retrieved and then converted into digital numeric values, which can be processed by the computer. The data acquisition system usually changes the analogue waveform into a digital value to further processing [6].

Figure 2. Data acquisition

The main equipment used in this study was a 30.5 cc backpack grass trimmer, which is STHIL. FR. 3000 model, with the size of 345x290x401 in millimetres and a weight of 9.4 kg, The damper materials used was in sheet type with a thickness of 10mm each. The materials are Natural Rubber, Silicone Rubber, Foam EVA (Ethylene-vinyl acetate). Each of sheet type materials is wrapped on the handle and bound.

The vibration sensor used in this study is the PIEZOTRONICS PCB production accelerometer sensor with a 352c33 model, and for signal reading it is used the Dynamic Signal Acquisition NI 9250 with BCN.

In this study, data collection was done on the handle of a grass trimmer, where the sensor is an accelerometer in the paste directly on the handle

3. Result and discussion

The focus of this research is the damping materials that the material was tested under vibration. The test was carried out twice with two different vibration sources but the dimensions and materials alike.

In the first test, the materials were tested by using a vibration exciter with the vibration acceleration of 0.3092 m/s2. The setup can be seen in Fig. 3. The test result shows that the three materials have a good performance in reducing vibration acceleration as can be seen in Table 1. However, Foam EVA has a better performance compared with two other materials.

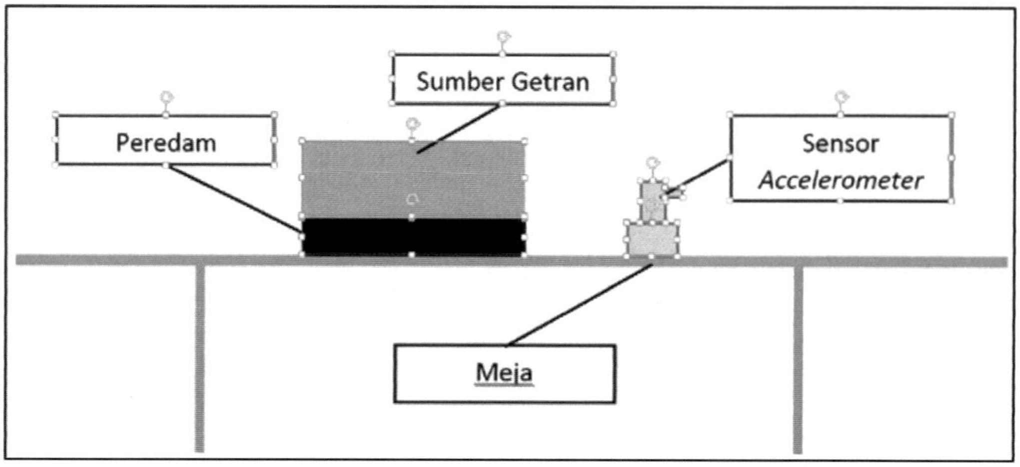

Figure 3. Setup of the vibration system by using vibration exciter

Table 1. Percentage of Acceleration Reduction when The System Vibrates at The Acceleration of 0.3092 m/s^2

Damper	Acceleration [m/s^2]	Percentage of Acceleration Reduction [%]
Natural Rubber	0.1362	56
Rubber Silicone	0.0860	72
Foam EVA (*Ethylene-venyl acetate)*	0.0446	85

In the second test, the material sheet was layered around the grass trimmer handle and an accelerometer was set above it to investigate the vibration acceleration. The result is shown in table 2. From the table, it can be seen that the Foam EVA has the best performance. Compared with the two other materials.

Table 2. Percentage of Acceleration Reduction when applied on The Grass Trimmer Handle

Damper	Acceleration [m/s^2]	Percentage of Acceleration Reduction [%]
Handle	23.56	0
Natural Rubber	18.27	22
Rubber Silicone	16.24	31
Foam EVA	14.80	37

From these two tests, we can see that the result shows a similar trend as can be seen in Figure 4. The graph shows that Foam EVA has the best performance in reducing the vibration acceleration, and Rubber Silicon and Natural Rubber has the second and third respectively.

From Fig. 4 we also can see the comparative level of vibration acceleration absorption ability of each damper material. The higher the vibration acceleration, the lower the acceleration reduction of the material, but the ratio of acceleration reduction between each material is not much different. Here it is found that EVA foam has the highest vibration acceleration reduction ability

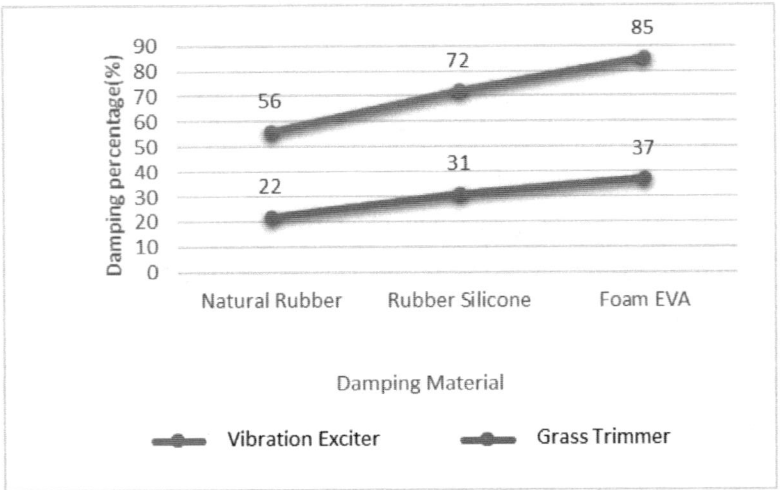

Figure 4. Comparison graph of vibration reduction between each damper with different vibration magnitude

4. Conclusion

The three types of damping material have good performance in reducing vibration acceleration on grass trimmer, which in this study EVA foam is the best compared with the other two dampers.

The higher the vibration acceleration, the lower the vibration acceleration reduction of the material, but the ratio of acceleration reduction between the material is not much different.

5. Reference

[1] Hasanuddin I Syahriza and Yulistia S 2017 Analisa Pengaruh Mesin Potong Rumput pada Badan Manusia Ditinjau dari Aspek Ergonomi Desain *Proceeding of SNTTM XVI*

[2] Hao K Y Ean O L and Ripin Z M 2011 The Design and Development of Suspended Handles for Reducing Hand-arm Vibration in Petrol Driven Trimmer, *International Journal of Industrial Ergonomics* **41** 1 469

[3] Zainuddin 2016 Desain dan Simulasi Peredam Getaran pada Dua Handle Mesin Potong Rumput Jenis Backpack Type FR3000 A Thesis Universitas Syiah Kuala

[4] Yoshida J Uemura M Miyakawa S Oono T and Ishikawa D 2013 Reduction of High Frequency Vibration of Brush Cutter by Structure Optimization *Proceedings of the World Congress on Engineering*

[5] Riyadi M Wahyudi and Setiawan I 2010 Pendeteksi Posisi Menggunakan Sensor Accelerometer MMA7260Q Berbasis Mikrokontrol Atmega32 *TRANSMISI* **12** 2 76

[6] Nataliana D Albuyani U A Parawansa M S A and Darlis A R 2016 Simulasi Data Acquisition Alat Uji Flight Control Actuator Pesawat Menggunakan Software Labview *Jurnal Elektro Telekomunikasi Terapan* **3** 1 280

IOP Conf. Series: Materials Science and Engineering **931** (2020) 012007 doi:10.1088/1757-899X/931/1/012007

Power density comparison of multi-air gap axial flux permanent magnet motor for electric vehicle

M Tadjuddin[1*], I Hasanuddin[1], Z Fuadi[1], A Mukminin[2], M R Arief[1]
Department of Mechanical and Industrial Engineering, Universitas Syiah Kuala, Darussalam, Banda Aceh, Indonesia
Department of Electrical Engineering, Universitas Syiah Kuala, Darussalam, Banda Aceh, Indonesia

E-mail: m.tadjuddin@unsyiah.ac.id

Abstract. Electric vehicles are a necessity now, especially in the future. The efficiency of electric vehicles continues to be improved, especially in the drive system or electric motor. One type of electric motor used in an electric vehicle that has high efficiency and power density is permanent magnet axial flux motor. This paper explains the comparison of power density on two designs of axial flux motor. A single and multi-air gap axial flux motors had been compared. The finite element method (FEM) simulation is used to determine the power and volume of those motors. The results presented that the two air gaps power density axial flux motors were 195,3 W/kg while those multi air-gap motor with four air gaps had 283.6 W/kg. The results shoes that four air-gap has a bigger power density than two air-gaps.

1. Introduction

Fuel oil from 2000 to 2014 total consumption increased to 315 million barrels of oil equivalent, along with the improving economic growth in Indonesia, this data is presented in Outlook Energy Indonesia 2016 by the Technology Assessment and Application Agency (BPPT). The pattern of fuel consumption in Indonesia, which is increasing every year, is one of the consequences of the Indonesian people's habit of increasing consumption levels of gasoline vehicles. Excessive use of fuel energy can be reduced by using electric vehicles because electric vehicles are one of the excellent and practical solutions and make these electric vehicles a necessity in the future.

Electric vehicles are based on an electric power system as all resources. There are no internal or external combustion engines onboard. Electric power as an energy source that can drive two-wheeled vehicles and four wheels. The main advantage is the high power efficiency conversion through the electric motor proposition system [1]. The efficiency of the motor is highly prioritized in designing an electric vehicle. One type of motor that shows excellent performance is the axial flux motor.

Axial flux motor type is also called motor with disk type. It is an attractive alternative to conventional motors because of its disk-likes shapes, compact fabrication, and high power density. Axial flux motors are most comfortable for electric vehicles, fans, pumps, valve controls, flight systems, machine tools, robots, and industrial equipment. This type of motor has a larger diameter than conventional motors. Thus providing high inertia that can be driven as a flywheel [2]. Motor efficiency can be measured using several software models. One of the factors to achieve flux density efficiency can be measured using FEMM software. This software is used and focused on analysing the magnetic flux density. To analyse the magnetic field working in the motor [3].

IOP Conf. Series: Materials Science and Engineering **931** (2020) 012007 doi:10.1088/1757-899X/931/1/012007

2. Theoretical background

2.1. Axial flux motor

Nowadays, many areas need more things like power density, low weight, low material, low cost, excellent cooling, and ventilation. Axial flux motors or often abbreviated as AFPM are motors that have good criteria to be used in the future. AFPM is more compact than the RFPM engine shown in Figure 1; it is recognized as having the ability to get more power density than an RF engine [4].

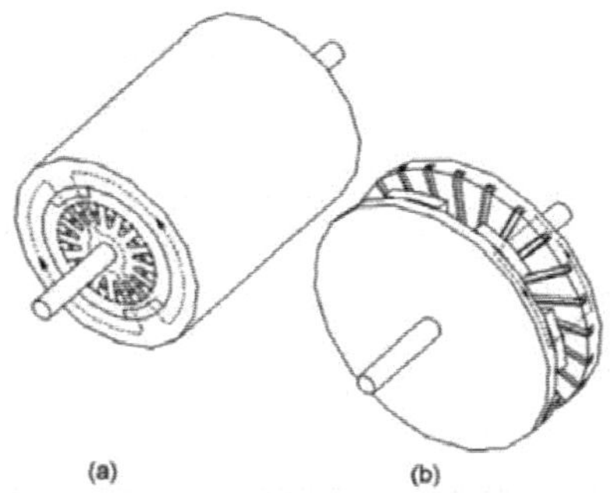

Figure 1. (a) RFPM engine, (b) AFPM engine [4]

AFPM machines are recognized that can get better power density than RFPM machines. Therefore AFPM can be applied because it is efficient and effective, and its capabilities are superior to RFPM [5]

2.2. Multi-sided axial flux motor

To assess the potential of a multi-stator engine, a double stator AFPM engine is considered. This machine has two disks (stator) that support entanglement (Figure 2). On the rotor, parts are connected three magnetic disks which are connected to each other in order to act as a driving shaft in the drive motor system which will later be connected to the wheels.

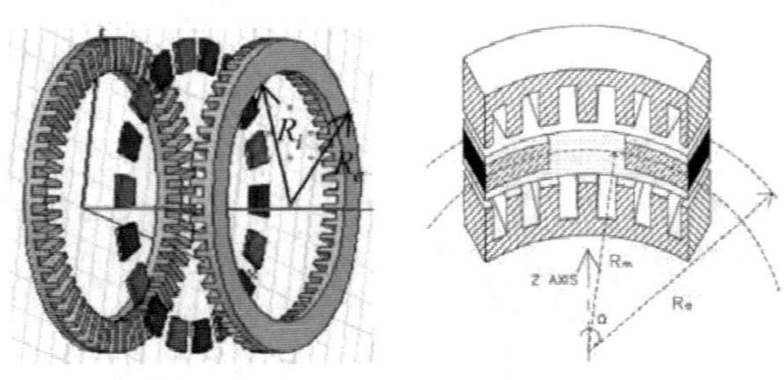

(a) (b)

Figure 2. Display (a) 3D dual stator AFPM engine, (b) 3D sketch pf a double stator AFPM machine [6]

2.2.1. Axial flux motor multi-sided design

In this paper, the axial flux motor is designed with a multi-sided model that has two winding disks, which become the stator while three interconnected magnetic disks which become the rotor Figure 2.

To develop an axial flux motor requires an initial step by determining the parameters as targets to be achieved in Table 1.

Table 1. Parameter motor flux axial

Parameter	Value	Unit
WxTxH Magnet Specifications	35x8x50	Mm
Type of Magnet	N52	-
Modules Magnet	32	Units
Inter Magnet Spacing	10	mm
Number of Twists	3	Turn
Diameter wire	.8	mm
Air gap	1	mm
Power	1000	Watt
Operational voltage	60	V
Number of Phases	3	Phase

The multi-sided flux axial motor design in this paper is designed with a coreless type (figure 3), where there is no metal element on the winding disk. Still, here the windings are coated with resin material. So there is no slot winding (slot less). The coreless type will not produce cogging torque so that the efficient level is better and very useful for use in electric vehicles

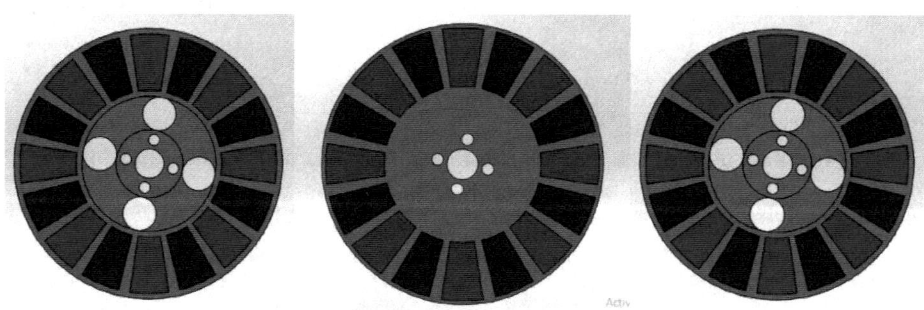

(a)

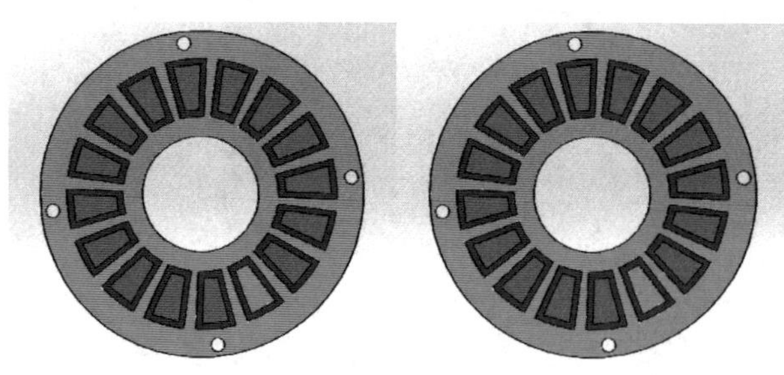

(b)

Figure 3. Display (a) three magnetic disks that function as rotors (b) two winding discs that function as stators

Figure 4. Design of multi-sided flux axial motors

To assume the maximum power density with no load, the thickness of the rotor seat can be calculated by the following equation:

$$y_r = \frac{1}{B_{satr}} \int_0^{\frac{\pi}{2p}} B_{rII}(R_r, \theta) R_r \, d\theta \quad (1)$$

Copper loss is calculated using the following equation:

$$P_{Cu} = 2I^2 R \quad (2)$$

On the stator stand the angle changes β depending on the angle θ by following the following equation:

$$\beta = \theta + \theta_0 + \omega t \quad (3)$$

2.2.2. Finite element method magnetics simulation

The simulation used in this paper is Software Finite Element Method Magnetics version 4.2 (FEMM 4.2). Figure 5. shows the topology of the axial flux motor in 2D, there are two sides of back iron with a thickness of 10 mm, on each side there are 4 magnets, the type of magnet used is NdFeb 40, with a water gap of 1 mm.

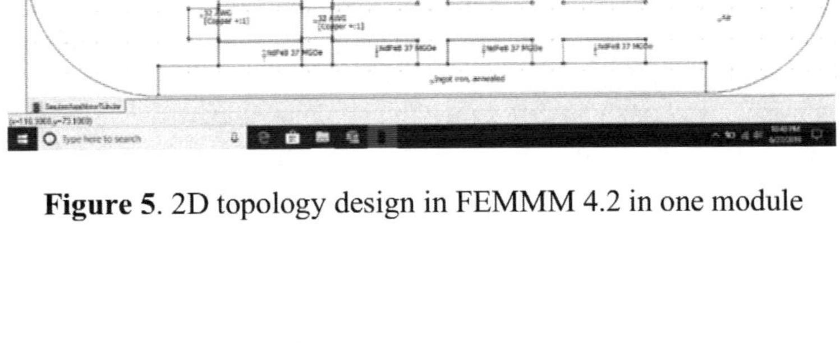

Figure 5. 2D topology design in FEMMM 4.2 in one module

3. Results and discussion

The simulation results show the flux velocity, the magnetic field lines can be measured for strength, and the magnetic flux density distribution in an axial flux motor module can be seen in Figure 6.

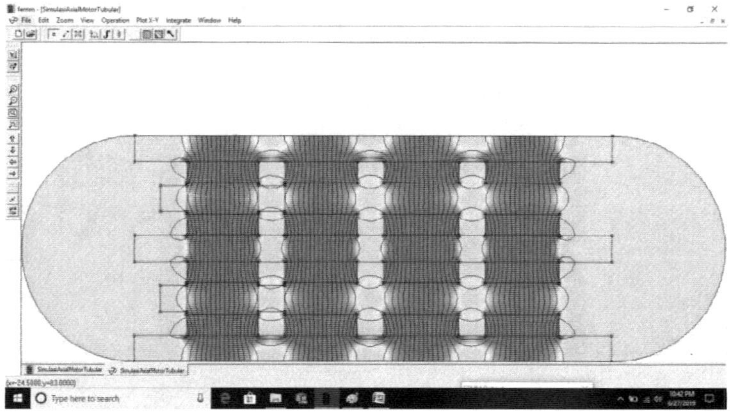

Figure 6. Distribution of flux in a simulation

Figure 6. shows the flux density, the flux density of each part of the axial flux motor in the simulation has a certain colour variation, where the colour explains and identifies the flux density.

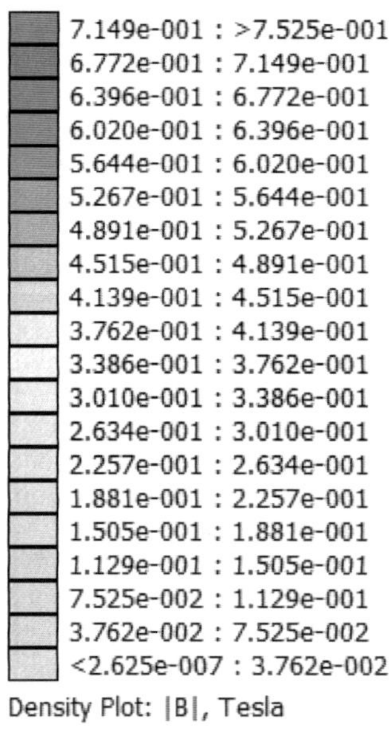

Figure 7. Legend flux density during simulation

It can be seen from Figure 7 that the distribution is good at the active part of the conductor. The maximum magnitude of the magnetic flux in the region also determines the triggering winding time. The amount of flux density in the region can be seen in the following figure.

The power output can be calculated from the flux density plotted above. The total weight of the motor is taken from 3D design properties.

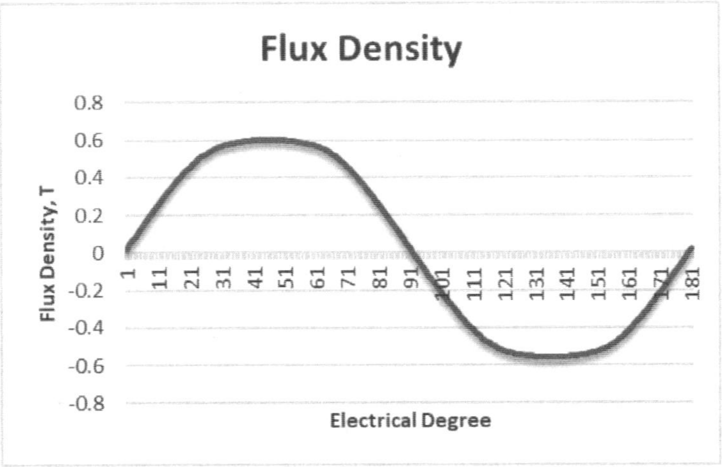

Figure 8. Flux density on the design, at one electrical degree

4. Conclusion

Power density means the amount of power per unit volume. The finite element method (FEM) simulation is used to determine the power and volume of multi-sided axial flux motors. The results showed that the two air gaps power density axial flux motors were 195,3 W/kg, while those with four air gaps motors had 283.6 W/kg. The results shoes that four air-gap has a bigger power density than two air-gaps.

5. Acknowledgment

This research is funded by RISTEKDIKTI on research grant of Penelitian Terapan 2019, under LPPM and fully support of the Laboratory of Design and Manufacture, Universitas Syiah Kuala.

6. References

[1] K. W. E. Cheng, "Recent development on electric vehicles," Power Electron. Syst. Appl. 2009. PESA 2009. 3rd Int. Conf., no. June 2009, pp. 1–5, 2009.
[2] C. Paper and F. D. Erciyes, "Prototyping of Axial Flux Permanent Magnet Motors," no. June 2015, pp. 1–7, 2016.
[3] N. Rahim, H. Ping, and M. Tadjuddin, "Design of an In-Wheel Axial Flux Brushless DC Motor for Electric Vehicle," in 2006 International Forum on Strategic Technology, 2006, pp. 16–19.
[4] O. S. Daif, "Axial Flux Permanent Magnet brushless machine, a new topology of electrical machines and brief about it," vol. 7, no. 11, pp. 1240–1248, 2016.
[5] R.-J. Wang, "Design Aspects and Optimisation of an Axial Fielf Permanent Magnet Machine with an Ironless Stator," no. March, 2003.
[6] S. Djebarri, J. F. Charpentier, F. Sculler, and M. Benbouzid, "Design and Performance Analysis of Double Stator Axial Flux PM Generator for Rim Driven Marine Current Turbines," IEEE J. Ocean. Eng., vol. 41, no. 1, pp. 50–66, 2016

IOP Conf. Series: Materials Science and Engineering **931** (2020) 012009 doi:10.1088/1757-899X/931/1/012009

A Study of Place Attachment to Mosques as an Evacuation Shelter Building

L Vivita[1], Husaini[2], R Anggraini[3], C Dewi[4]
[1]Student of Engineering Doctoral Program, Universitas Syiah Kuala, Darussalam, Banda Aceh, Indonesia
[2]Mechaninal and Industrial Engineering Department, Universitas Syiah Kuala, Darussalam, Banda Aceh, Indonesia
[3]Civil Engineering Department, Universitas Syiah Kuala, Darussalam, Banda Aceh, Indonesia
[4]Architecture Engineering Department, Universitas Syiah Kuala, Darussalam, Banda Aceh, Indonesia

E-mail: lolavivita@yahoo.com

Abstract. The city of Banda Aceh has a geological location that is prone to earthquakes and tsunamis. In 2004, the tsunami damaged the city structure, and many victims died. When the tsunami struck the area, a mosque was used as an evacuation building. Its scatter existence in residential areas made it easy to reach, but several mosques could not be used as an evacuation building. The purpose of this study was to explore the function of mosques as tsunami evacuation building through place attachment studies using qualitative methods and depth interviews to identify public attachment factors. To measure place attachment, this research used three variables, i.e. socio-demographic, environmental experience, and familiarity with the place. This study also used the space forming variables, which include meaning, function, and form to see the public attachment to a mosque as an evacuation shelter building. The results showed that place attachment to a mosque was influenced by its meaning as a sacred place. The majority of people in Banda Aceh are Muslim and have Islamic cultural backgrounds. People's routine activities form strong bonds to a mosque. Acehnese culture to glorify guests is reflected in mosque building design by exploring open walls, ease of accessibility, and spread presence in residential areas. These cultural characteristics strengthen the opportunity to make mosques as tsunami evacuation shelter buildings in Banda Aceh.

1. Introduction

Banda Aceh is the capital of the Indonesian Province of Aceh with an area of +61.36 km^2. Geologically, Banda Aceh at the meeting of the Euroasia Plate and Australia is ± 130 km from the western coastline, so it is prone to tsunamis (Spatial Plan of Banda Aceh City 2009-2029). On December 26, 2004 at 07.58 of West Indonesian Time, Banda Aceh suffered an earthquake with a magnitude of 9.3, which caused a tsunami as high as 10-15m and resulted in the damage of the most of territory [1], [2] with fatalities of ± 160,000 people [3]. On April 11, 2012 an earthquake measuring 8.6 on the Richter scale struck the city. Although it did not cause tsunami, people still evacuated to safety place (high land and building), when the tsunami struck, the mosque was used as an evacuation shelter building, this was confirmed by some researchers [4] an [5].

The mosques spread in residential areas makes them easy to reach, although some mosques cannot be used as evacuation shelter buildings, especially in areas with high tsunami threat potential such as in Meuraxa and Kuta Raja Districts. The function of mosque in Banda Aceh is as a meeting place/deliberation, social activities, the celebration of Islamic holidays, study / dayah, preaching, and protection, so automatically, people have a strong attachment to the mosque. Attachment is the emotional impact of a place where people are attracted to emotional and cultural ties [7]. Site attachment is one of the dimensions of sensitivity to place and positive emotional bonds that develop between place and individual [8]. Attachment to a place is the inner bond of a person or society to a specific place [9]. This inner bond can bring emotional feelings and functional dependence on the place. Feelings of dependence will be awakened if someone knows them well and feels that the place is very meaningful [10]. Experience and memories will create meaning in place [11]. He explained that not only the place itself is significant, but the experience of the place can also create meaning. The meaning is often related to 3 (three) important things, namely self, other people, and the meaning of the environment itself [12].

Aside from being a place of worship, a mosque as a public building is also used as an evacuation shelter. Public buildings that have met the structure and construction requirements can be used as evacuation protection buildings. This additional function confirms that each evacuation protection building is a multi-functional building [4]. Buildings to be used as evacuation shelters must have a public function or a function-oriented to public services. Examples are mosques, schools [1], hotels, restaurants, government offices, parliament buildings [6], conference center buildings, shopping centers [1], sports centers, parking lots, and markets. The purpose of this study was to explore the attachment of a place to the mosque as a tsunami evacuation protection building in Banda Aceh.

2. Method

2.1. Place Attachment

This study uses qualitative methods through direct interviews to identify the factors of community attachment to mosques, especially its function as an evacuation shelter building. The selection of respondents was carried out randomly without having to determine the object and place of the interview, so that information gathering was obtained from various levels of society. Interview questions include:

1. In the case of a tsunami disaster, what evacuation building is chosen? Vertically or horizontally? why?
2. If vertical evacuation is carried out, what function of the building will be chosen and why?
3. When choosing a mosque as a vertical evacuation, what factors make a mosque suitable and comfortable to use as an evacuation building?
4. With regard to the phenomenon of several mosques surviving during the 2004 tsunami, what factors do you think caused the mosque to survive?
5. Is there anything else that can strengthen the function of the mosque as an evacuation site other than the factors of meaning, function and form of the mosque itself?

The five questions above was broken down into other questions while the interview process was underway. The purpose was to extract detail information from the respondents so that important data could be obtained in this study. To measure place attachment, the variables used are socio-demographic, environmental experience, and familiarity with the place. This is explained by [13] in research on the attachment of places, emotional feelings, and functional feelings were used to measure the level of one's attachment to a place. The attachment of a place is influenced by:

1. Socio-demographic characteristics [14]; [12]; [15];
2. Environmental experience, including the type of involvement of people with place [15];
3. Degree of familiarity with places [16];
4. Culture [17];
5. The place itself.

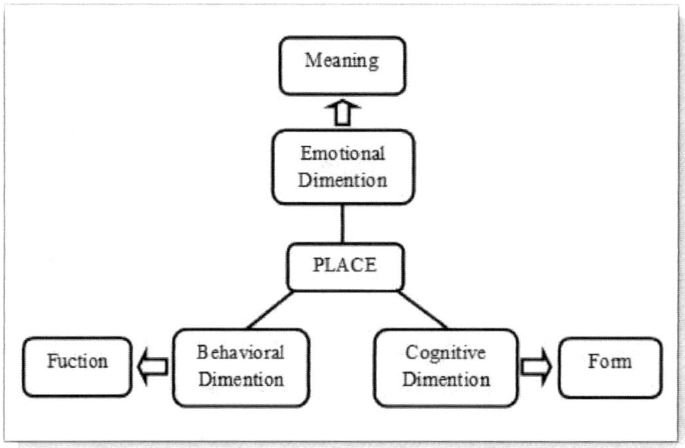

Figure 1. The dimension of place forming.

A place is a space that is related to human relations to the physical environment, individual and group activities, concept or meaning [19], [20] as well as cultural processes [14]. Space and place are interconnected, where the physical aspects of space are considered important based on the values given by humans to the place [21]. The place can provide comfort for its users or even vice versa. The comfort of a place, both physically and psychologically, is one of the important elements for humans to carry out activities in a physical environment. Comfort will trigger a person's feelings towards the atmosphere they experience in carrying out their activities [22]. Differences in the characteristics of communities with respect to places have their own concept of attachment. The physical characteristics of the environment do not only distinguish places, but they also affect people's perceptions of meaning [7]. This is explained by [23] who states that the concept of attachment to a place in eastern cultures, especially Indonesia, must consider local perspectives so that researchers can understand the local characteristics of natives in forming their attachment to the place.

The meaning of a place is a concept that transforms space into a place with special behavior and sensory characteristics of certain people [7]. Spatial perception arises because of interactions in the cognitive dimension (form) so that environmental elements can be known and used as directions. Interactions on the dimensions of behavior (function) are used for activities and functional relationships between humans and the environment. The meaning of a place (meaning) has descriptive and emotional aspects of environmental experience, which means that the concept of a place meaning is a psychological and physical concept [7]. Physical parameters that affect the meaning of a place are size, scale, component, diversity, texture, decor, color, smell, noise, and temperature. The way people communicate with places is influenced by identity, history, happiness, mystery, convenience, beauty, security, vitality, and memories [24]. Place identity is the concept of an individual's emotional attachment to a place or environment. Place identity is supported by physical dimensions and social environment [25]. A place can be identified if it can encourage people to stay longer and connect with one another. The identified place has a real picture, and people know where and when to be there [26]. This is related to the ability of the place to awaken the human senses through quality so as to make it different from others [27].

The concept of aplace identity basically reviews the relationship between who and where we are and how our local environment (including geographical location, cultural traditions, cultural heritage, etc.) affects our lives [28]. In Banda Aceh, the majority of the population is Muslim with inherent Islamic cultural traditions, so that a mosque is widely used in social activities.

2.2. Requirements of earthquake and tsunami resistant building

With its function as a tsunami protection building, a mosque must meet the requirements of earthquake and tsunami resistant buildings as determined by the Directorate General of Human Settlements on Technical Guidelines for Houses and Buildings that are Earthquake Resistant Buildings:

1. Located on stable ground;
2. Has a simple and symmetrical plan;
3. Continuous foundation without interruption;
4. The area of the wall 12 m2 mounted column;
5. The circular beam is bound rigidly with a column;
6. Building framework is firm and rigid;
7. Using dry wood as a construction easel;
8. Light wall material and fastened column;
9. Material of brick / brick wall is not broken and sounds loud when mixed;
10. Correct composition of species / mix;
11. Has a bond between components (structural / non-structural).

Evacuation protection buildings are defined as buildings that function as tsunami evacuation destinations. In some literature, they are also referred to as rescue buildings [1], [29], and vertical protection buildings [6].This study also used the Spatial Dimensions of Meaning, Function, and Form to see the community's attachment to a mosque as an evacuation protection building. This is as explained by [18] where the theory of "attitude / behavior" defines three dimensions of place, namely the meaning of place (emotional dimension); the shape of place (cognitive dimension); and place function (behavioral dimension) [18]. Place dimension is shown in Figure 1.

2.3. Requirements of evacuation shelter building

An important point in determining evacuation protection buildings is that buildings can withstand disasters and have a floor height above the level of a tsunami inundation. Evacuation shelter building requirements are:

1. The structure of the building is not damaged and can function as a temporary shelter;
2. Evacuation floors must be higher than wave height;
3. Its function can be attached to public buildings such as mosques, schools, meeting centers, shopping centers, sports centers, parking lots and markets [29], hotels, restaurants, government buildings, parliament buildings [6];
4. Design and capacity that can accommodate many people;
5. Location or ease of access horizontally;
6. Easy access vertically;
7. Good security.

3. Results and discussion

Based on field interviews, community attachment to mosques was influenced by differences in age, environmental experience, and familiarity with places. The number of respondents was 20 people and resided in Banda Aceh. The selection of respondents was random to obtain various information. Respondents were between 20 - 70 years of age, and they were lecturers, doctors, college students, housewives, civil servants and private employees with a percentage of 60% women and 40% men. Table 1 shows the public attachment to a mosque as a tsunami evacuation shelter building in Banda Aceh.

Table 1. Public attachment to a mosque as an evacuation shelter building.

Variables of Place Attachment	Public Attachment to a Mosque as an Evacuation Shelter Building
Socio-Demography (Age-Years): $20 \leq 30$	1. The hard and sturdy shape of a mosque giving the impression of security
$31 \leq 50$	1. The meaning of mosque as a sacred place 2. The hard and sturdy shape of a mosque giving the impression of security
≥ 50	1. The meaning of mosque as a sacred place 2. Accessibility to the first floor (place to prayer) is not too high, so easy to do the routine worship 3. Easy to reach from living home
Environmental Experience: Direct tsunami victims	1. The meaning of mosque as a sacred place and the history of survival mosque at tsunami disaster 2. Available accessibility to the upper floor 3. Easy to reach from living home 4. The building façade is unique and easily recognizable
Indirect tsunami victims	1. The hard and sturdy shape of a mosque giving the impression of security 2. Easy to reach from living home
Familiarity With The Place: Frequent activities at a mosque	1. The meaning of mosque as a sacred place 2. Worship consistency brings intimacy to a mosque
Rarely activities at a mosque	1. The hard and sturdy shape of a mosque giving the impression of security 2. Easy to reach from living home

This research used variables such as socio-demographic, environmental experience, and familiarity with place to see attachment of place where these three variables became indicators of public attachment in the mosque as a tsunami evacuation protection building based on spatial dimensions that included meaning, function and form. Respondents have experienced being directly and indirectly affected by the tsunami, where familiarity with the mosque was assessed based on whether or not they frequently carried out activities at the mosque. The interview results showed respondents aged $20 \leq 30$ years chose the mosque as an evacuation place because of the form factor which visually gave a high, sturdy and strong impression especially if the columns / pillars of the building could be properly exposed, while respondents aged $31 \geq 50$ years preferred the mosque because the meaning / value contained in the mosque as a sacred building. In addition, the ease of accessing both horizontally and vertically is also one of the considerations when they chose the mosque because of limitations in their motor systems. The experience of direct and indirect tsunami victims had a different attachment. Some chose the mosque because it meant a sacred place for them,

IOP Conf. Series: Materials Science and Engineering **931** (2020) 012009 doi:10.1088/1757-899X/931/1/012009

they were sure that being in the mosque would provide calm and security, and they were resigned themselves and sincere to all possibilities that would occur.

However, some respondents actually chose to evacuate horizontally by looking for a higher place such as mountains or hills, but if they had to evacuate to the mosque, the mosque must be high, strong, and sturdy. This is due to trauma from the events they experienced. Respondents who often performed religious activities in the mosque have a strong bond with the mosque, so they chose the mosque because it was a sacred place. Overall, the place attachment variable that greatly affected the community selection of a mosque as an evacuation place is "familiarity with the place." This is due to the religious activities carried out routinely and continuously in the mosque so that high faith evoked the meaning of the mosque as a sacred building that could provide comfort and safety in it.

4. Summary

Place attachment to a mosque as a tsunami evacuation shelter building was strongly influenced by its meaning as a sacred place because the majority of people in Banda Aceh are Muslim and have an Islamic cultural background. The worship routine activities of people created a very strong bond to a mosque. The tradition and social life of the people in Banda Aceh are those of the call to prayer during disasters such as the earthquake and tsunami. The behavior of people going to the mosque when the call to prayer is an opportunity to make the call to prayer as a siren so that people evacuate to the mosque because psychologically, the sound of the call to prayer can provide calm and comfort. Acehnese culture to glorify guests is reflected in mosque building design by exploring open walls, ease of accessibility, and spread presence in residential areas so that it is easy to reach when a disaster occurs. These cultural characteristics strengthen the opportunity to make mosque as a tsunami evacuation shelter building in Banda Aceh.

5. References

[1] JICA BAPPENAS Provincial Government of Nanggroe Aceh Darussalam 2005 The Study on the Urgent Rehabilitation and Reconstruction Support Program for Aceh Province and Affected Areas in North Sumatra

[2] Lavigne F, Paris R, Grancher D, Wassmer P, Brunstein D, Vautier F, Leone F, Flohic F, Coster B, Gunawan T, Gomez C, Setiawan A, and Cahyadi R 2009 Reconstruction of Tsunami Inland Propagation on December 26, 2004 in Banda Aceh, Indonesia, through field investigations *Pure Appl. Geophys 166* **259–281**

[3] Frankenberg E, Gillespie T, Preston S, Sikoki B, and ThomasD 2011 Mortality, the family and the Indian Ocean Tsunami *Econ. J. 121* **162–182**

[4] Budiarjo A 2006 *Evacuation Shelter Building Planning for Tsunami-prone Area; a Case Study of Meulaboh City, Indonesia* The Netherland International Institute For Geo-Information Science and Earth Observation Enschede **1-112**

[5] McCaughey J W, Mundir I, Daly P, Mahdi S and Patt A 2017 Trust and distrust of tsunami vertical evacuation buildings: Extending protection motivation theory to examine choices under social influence *submitted to Journal of Disaster Risk Reduction 24* **462-473**

[6] Eisner R. and NTHMP 2001 *Designing for Tsunamis Background Papers. US National Tsunami Hazard Mitigation Program Steering Committee* Sacramento **205**

[7] Hashemnezhad H, Heidari A A and Hoseini P M 2013 Sense of Place and Place Attachment *submitted to Journal of Architecture and Urban Development 3* **5-12**

[8] Steadman C R 2003 Is it really just a social construction: The contribution of the physical environment to sense of place *Society and Natural Resources 16* **671-685**

[9] Hidalgo M C and Hernandez B 2001 Place Attachment: Conceptual and Empirical Questions *submitted to Journal of Environmental Psychology 19* **331-352**

[10] Stokols D and Shumaker S A 1981 *People in places: A transactional view of settings. In Harvey, J. H. (Ed.). Cognition social behaviour and the environment* Hillsdal NJ. Lawrence Erlbaum Assoc.

IOP Conf. Series: Materials Science and Engineering **931** (2020) 012009 doi:10.1088/1757-899X/931/1/012009

[11] Manzo L C 2005 For better or worse: Exploring multiple dimensions of place meanin: *submitted to Journal of Environmental Psychology 25* **67–86**

[12] Gustafson P 2001 Meanings of Place: Everyday Experience and Theoretical Conceptualizations *submitted to Journal of Environmental Psychology 21 (1)* **5–16** and Roots and Routes Exploring the Relationship between Place Attachmentand Mobility *Environment and Behavior 33(5)* **667–86**

[13] Heryanto B, Ihsan and Natalia V V 2012 City identity and place attachment *submitted to Journal of Engineering Faculty 6* **3**

[14] Altman I and Low S M 1992 *Place attachment* New York Plenum Press

[15] Williams D R and Roggenbuck J W 1989 *Measuring place attachment: Some preliminary results. Paper presented at the Outdoor Planning and Management NRPA Symposium on Leisure Research* San Antonio TX.

[16] Fried M 1963 *Grieving for a lost home, in The urban condition: people and policy in the Metropolis* Duhl J L Editor New York Basic Books, Inc. **151-171**

[17] Tuan Y F 1974 *Topophilia: A study of environmental perception, attitudes, and values* Englewood Cliffs NJ Prentice-Hall

[18] Jorgensen B S and Stedman R C 2001 Sense of place as an attitude: Lakeshore owners attitudes toward their properties *submitted to Journal of Environmental Psychology 21* **233-248**

[19] Canter D 1977 *The psychology of place* The Architectural Press Ltd.

[20] Smaldone D, Harris C and Sanyal N 2005 An exploration of place as a process: The case of Jackson Hole, WY *submitted to Journal of Environmental Psychology 25* **397-414**

[21] Zakariya K, Mohyuddin A, and Yaman M 2007 Refining tourist's place experience through placemaking: Concepts and correlations *submitted to Journal of Diversity in Organisations, Communities & Nations 4* **249-257**

[22] Lynch K 1960 *The image of the city* Mass MIT Press

[23] Sudrajat I 2012 Conceptualizing a framework for research on place in Indonesia. *SAPPK. ITB* (unpublished)

[24] Steele F 1981 *The sense of place* CBI Publishing Company, Inc.

[25] Twigger-Ross C L and Uzzell D L 1996 Place and identity processes *submitted to Journal of Environmental Psychology 16* **205–220**

[26] Yuen B 2003 *Searching for place identity in Singapore* Habitat International 29 **197-214**

[27] Shamsuddin S and Ujang N 2008 *Making places: The role of attachment in creating the sense of place for traditional streets in Malaysia*, Habitat International 32 **399-409**

[28] Fisher J J 2006 Creating Place Identity: It's Part of Human Nature, Course Description of Place, Identity and Difference, *Built Environment Geography*

[29] Bappenas 2005 Master Plan for the Rehabilitation and Reconstruction of the Regions and Communities of the Province of Nanggroe Aceh Darussalam and the Island of Nias, Province of North Sumatera

IOP Conf. Series: Materials Science and Engineering **931** (2020) 012010 doi:10.1088/1757-899X/931/1/012010

Analytical study vessel design of the portable patchouli oil steam boiler based on vacuum technology

A Razali, A Z Mubarak, M Alkindi, N H Aprilia, I Idram
Departmen of Mechanical and Industrial Engineering, Universitas Syiah Kuala, Darussalam, Banda Aceh

E-mail: asbar@unsyiah.ac.id

Abstract. Aceh geography is suitable for growing pogostemon cablin with high quality. Traditionally, patchouli oil extraction is carried out using distillation techniques. However, the results are unsatisfactory due to impurities substances presence in the boilers. High energy wastage is common in the traditional distillation technique process due to improper vessel design. The study aims to design and rate the vessel of portable distillate patchouli oil based on vacuum technology. The simplified vessel design and proposed vessel design were then tasted to find a proper design with respect to ASME boilers and pressure-related VIII Vessel Code. However, internal and external pressure employed in both design and operation, given that the simplified model is 56.91 MPa and S_a is 116 MPa. The vessel design is yield to proposed model stress (K_{mean}) $\leq S_a$ K_{mea} is 86.13 MPa and S_a is 116 MPa. The analytical calculation and simulation results show that von-mises stress is less than the allowable stress material.

1. Introduction

Aceh is a competitive producer of patchouli oil (pogostemon cablin, benth) containing patchouli alcohol (pa) with a 30% base quality standard [1]. Ministry of law and defense in the republic of Indonesia set up rules protecting Aceh patchouli plantation geographic region specifically with IDG 000000021 certificate [2]. In the world's market, patchouli oil use is in fragrance products. It is also popular as a raw material for a number of products such as antiseptics, aromatherapy, cosmetics, and a fixative to bind other essential oils. Its flavor is difficult to substitute with other synthetic compounds.

Traditionally, patchouli oil is extracted from the leaves, stem, and branches using the distillation techniques [3,4]. Currently, this technique is not optimally producing patchouli oil with high quality in view of the fact that the oil contains unwanted chemical substances such as Ferrum (Fe) and Cuprum (Cu). The impurities trigger the patchouli oil color, turning it dark causing PA level lower Indonesia National Standard (SNI). At koperasi KINA Aceh Jaya, Indonesia, most patchouli farmer groups apply the used-oil drum as a vessel that is a corrosive material and firewood as a source of energy to distillate patchouli oil. Distilled steam is only produced by a single unit of boiler from two units of the available. Steam is not properly distributed around of vessel due to lack of optimum design vessel. Extracting dry pogostemon cablin takes approximately 7-8 hours, which is not worth compared to end results. Therefore, the design and material type of a distillation boiler significantly contribute to low grade and economic values of patchouli oil. The objective of this study is to design and evaluate the vessel type of portable distilled patchouli oil.

2. Material and methods

2.1. Material

The patchouli leaves (pogostemon cablin, benth) used in this work, comes from Aceh Jaya, Aceh, Indonesia. The raw materials kept at room temperature in the storage facility. Experimental work carried out using distilled water.

2.2. Methods

The proposed design model of portable distilled patchouli oil vessel using vacuum technology refers to ASME VIII Boiler and Pressure Vessel Code requirement [5,6], as shown in Figure 1.

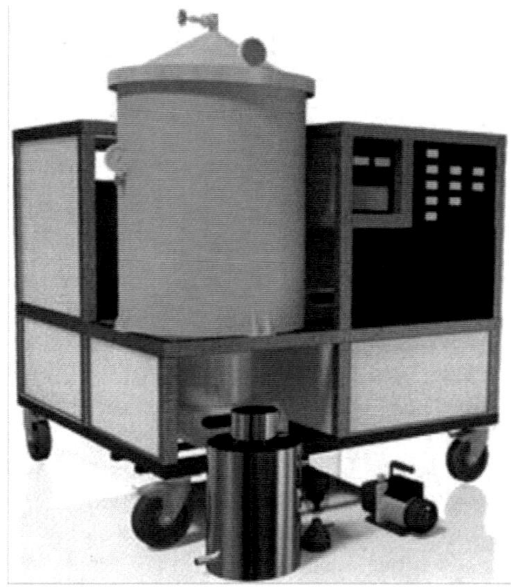

Figure 1. 3D model of portable distilled patchouli oil unit

To enhance the patchouli oil distilling process, six main issues were considered (as shown in Figure 2):

a) Energy source: Electric source was replaced with firewood to avoid contamination from hydrocarbon substances, and it was manageable by a control system.
b) Reliability, the system is a redesign to improve the performance of the current vessel and attain less negative impact on the environment (green technique) with a reasonable cost of production.
c) Purification water: The level of purification water surrounding the farm area was not appropriate enough to be used as raw water because it steamed in the distillation process. The purification water system was installed to provide water supply.
d) The pressure vessel was to consider design under the high-temperature condition and different energy sources in order to perform efficiency of energy and water utilities.
e) Condenser: Condenser pipe length was designed longer than the current system. Quick coolant system (QCS), with a cooling material temperature near 00C, was used to maximize the steam condensing process to condensate water.
f) Frame structure as the main board of the system was designed for an integrated portable system that was easy to carry and move from one place to another.

Pressure vessel construction design played a key role in the system. Its main parts consist of shell and head (see Fig.3). In the distillation process, the vessel worked under a certain number of pressure conditions. Since stress and strain worked in the vessel, a calculation was necessary to maintain life span and vessel reliability.

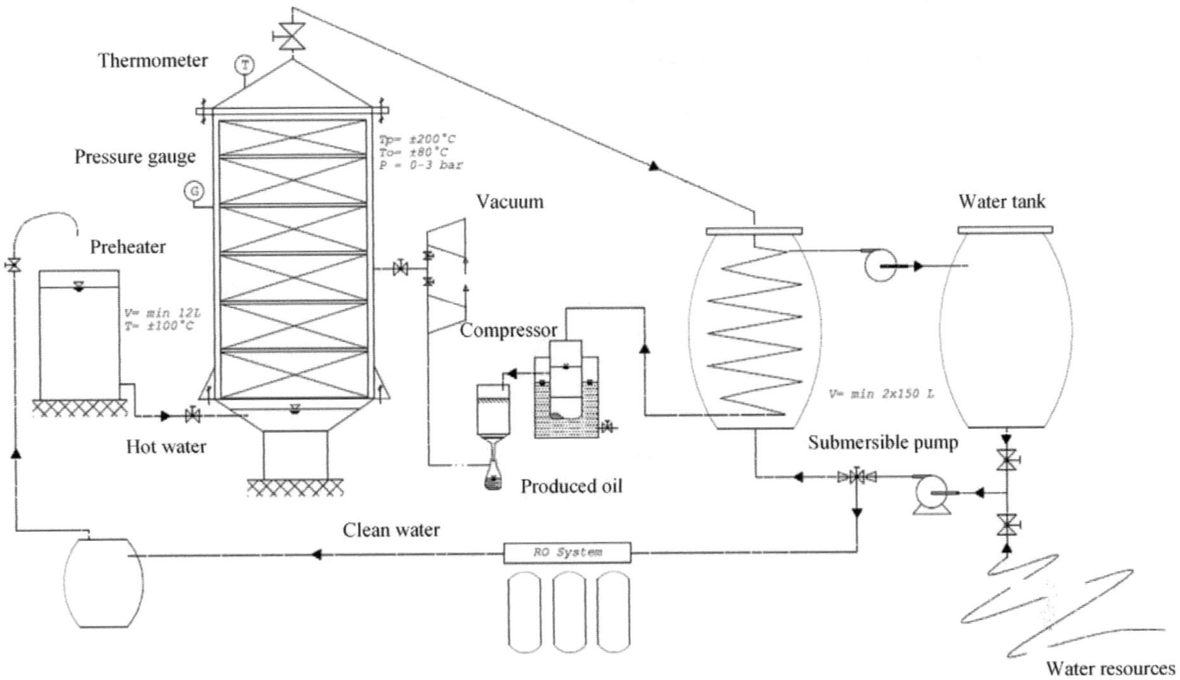

Figure 2. A schematic diagram of the portable distilled patchouli oil vessel using vacuum technology

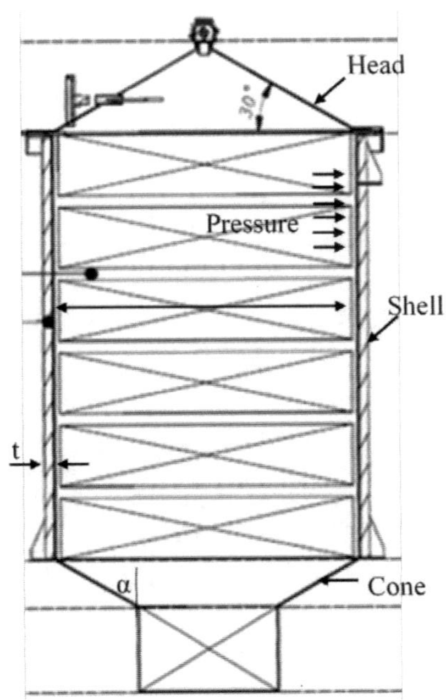

Figure 3. Vessel design and its main parts

In the shell section, the circumferential and latitudinal stress are in common existence and vice versa for head section meridional and latitudinal presented. This type of stress does not exist for the cone and conical shape of the vessel head. Stress works in shell calculated refer to relation in (P_d) pressure design (psi), (r_m) mean radium (in), (r_m). The thickness required of shell, head, or cone (in) is presented in the Eq. 1:

longitudinal stress

$$\sigma_{\emptyset} = \frac{Pr_m}{0,2\,t} \quad (1)$$

circumferential stress

$$\sigma_{\emptyset} = \frac{Pr_m}{t} \quad (2)$$

for cone and conical shape, working stress depends on the angle between two of cone diameters (α):

longitudinal stress

$$\sigma_x = \frac{Pr_m}{2 + \cos\alpha} \quad (3)$$

circumferential stress

$$\sigma_{\emptyset} = \frac{Pr_m}{t\,\cos x} \quad (4)$$

Distortion, stress equivalency or von-misses stress is reference stress that is commonly used to analyze and predict design failure. (σ_{vm}) Von-misses stress (psi) refers to (σ_x) longitudinal stress (psi), $\sigma_{\emptyset}$ circumferential stress (psi) and direction of welding (hoop), as present in Eq. 6:

$$\sigma_{vm} = \sqrt{\frac{(\sigma_{x\,tot} - \sigma_{\emptyset\,tot})^2 + (\sigma_{\emptyset\,tot} - \sigma_{t\,tot})^2 + (\sigma_{t\,tot} - \sigma_{x\,tot})^2}{2}} \quad (5)$$

In this study, a 3D model of vessel design and finite element analyses were performed on a desktop computer (IntelCore™ i7-4710HQ, 3.5 GHz processor, RAM 16 GB, Windows 10 operating system). The capacity of the vessel design proposed was 30 Kg and 20 litters of distilled water as suggested by farmers group KINA Aceh Jaya, Aceh, Indonesia. Detailed design of vessel parameters is shown in Table 1 and the material used in the computational simulation (stainless steel, thickness 1/8 in) is shown in Table 2.

Table 1. Vessel design condition and operation parameters

Initial pressure (operation) (Psi)	Initial pressure (vacuum) (Psi)	Operation pressure (Psi)	Designed pressure (Psi)	Temp. (F)	Designed temp. (F)	Heater temp. (F)
0	4351132	4351132	7251887	176	572	392

Table 2. The material of vessel design

Stainless Steel	Composition	Tensile (Psi)	Yield (Psi)	Min. Temp. (F)	Max. Temp. (F)	Corrosion factor	Welding factor
UNS. No. S30400	18Cr-8Ni	74984.51	30022.81	-427	1500.8	0.03	0.6

3. Result and discussion

According to the calculation results of longitudinal stress ($\sigma_{\emptyset}$) and circumferential stress (σ_x) in both process and vessel simulation were used to evaluate vessel design, as shown in Table 3. For welding joint area, the length of welding was calculated by considering the working stress at the vessel joint (293.4 in). Thereby, design stress (σ_t) was 1.977 psi and equivalent stress (σ_{vm}) was 9409.1 psi or 64.8 Mpa.

Table 3. Results of stress calculation for shell and head vessel design

	Components	σ_x ***(psi)***	$\sigma_{\emptyset}$ ***(psi)***
Shell	process	37725.53	7545.10
	vessel simulation	28968.69	5793.73
Head	cone	7545.10	15090.21
	conical	3450.27	6900.54
	flat	28968.69	0

The result of equivalent stress (σ_{vm}) from the model simulation was evaluated in design and operation respectively, as shown in Fig. 5. The material strength at temperature design (S_a) was 116 Mpa.

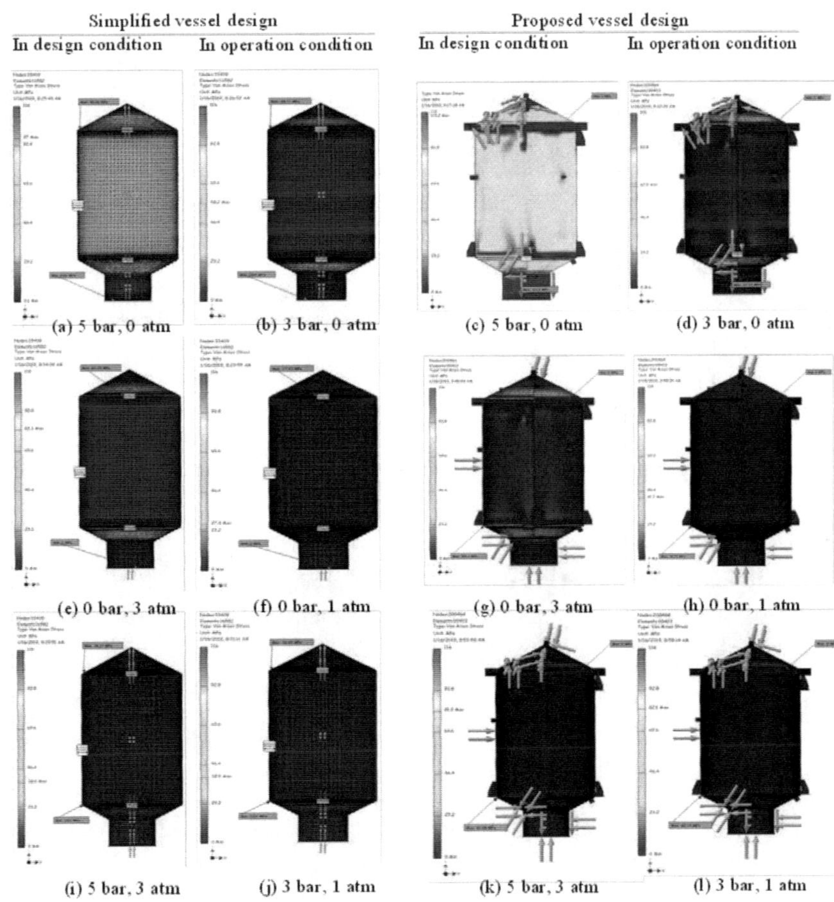

Figure 4. Simulation results of vessel design for two types model, (a), (b), (e), (f), (i), (j) simplified design and (c), (d), (g), (h), (k), (l) proposed design model in design and operation condition

In the computation, simulation simplified model of vessel and proposed vessel model were compared. The vessel model was evaluated in two conditions, in design and operation. In design condition, the loading parameter for the simplified model was performed either at internal pressure (5 bar) or external pressure (3 bar) as well in both combination internal and external pressures, as shown in Fig. 4 (a), (b), (e), (f), (i), and (j). In operation condition, the loading parameter for the proposed model was performed either at internal pressure (3 bar) or external pressure (1 atm) as well as a combination in both internal and external pressures, as shown in Fig. 5 (c), (d), (g), (h), (k), and (l).

For the analytical calculation of vessel design $\sigma_{vm} \leq S_a$, the result was σ_{vm} = 64.8 MPa, and allowable stress material (S_a) in design was 116 MPa. Meanwhile, the model simulation result was required to meet simplified model stress (M_{mean}) $\leq S_a$, M_{mean} of 56.91 MPa and S_a of 116 MPa (Table 4). The vessel design was obtained to proposed model stress (K_{mean}) $\leq S_a$ K_{mea} of 86.13 MPa and S_a is 116 MPa (Table 5). It can be concluded that the proposed model is acceptable to be manufactured.

Table 4. Results of pressure calculation for simplified model design

Pressure load	Condition	Pressure (MPa)	Highest pressure (MPa)	Lowest pressure (MPa)	Average (K_{mean})	S_a (MPa)
5 bar ; 0 atm	Internal pressure design	96.96	96.96	27.43	56.91	116
3 bar ; 0 atm	Internal pressure operation	58.17				
0 bar ; 3 atm	Internal pressure design	82.29				
0 bar ; 1 atm	Internal pressure operation	27.43				
5 bar ; 3 atm	Design pressure	38.27				
5 bar ; 1 atm	Actual operation pressure	38.39				

Table 5. Results of pressure calculation for the proposed model design

Pressure load	Condition	Pressure (MPa)	Highest pressure (MPa)	Lowest pressure (MPa)	Average (K_{mean})	S_a (MPa)
5 bar ; 0 atm	Internal pressure design	113.20	125.2	41.72	86.13	116
3 bar ; 0 atm	Internal pressure operation	67.93				
0 bar ; 3 atm	Internal pressure design	125.2				
0 bar ; 1 atm	Internal pressure operation	41.72				
5 bar ; 3 atm	Design pressure	81.28				
5 bar ; 1 atm	Actual operation pressure	82.10				

4. Conclusion

The portable vessel design of distilled patchouli oil based on vacuum technology has been designed and evaluated. It was developed to substitute the current design in order to increase the reliability and the quality of the final distilled oil products. Calculation and simulation for proposed vessel design should meet the allowable standard. For analytical calculation *of* $\sigma_{vm} \leq S_a$, σ_{vm} was 64,87 Mpa, and allowable stress material (S_a) in the design condition was 116 Mpa. Therefore, the proposed vessel design was more appropriate to apply in a portable patchouli oil steam boiler and suitable for manufacturing stages.

IOP Conf. Series: Materials Science and Engineering **931** (2020) 012010 doi:10.1088/1757-899X/931/1/012010

5. Reference

[1] Directorate General of National Export Development 2015 Ministry of Trade of The Republic of Indonesia *Export News Patchouli Oil* ed. Sugiarti Ditjen PEN/MJL/68/IX/2015
[2] Dahlan S 2013 Buku Persyaratan Indikasi Geografis Minyak Nilam Aceh [The Handbook of Geographic Requirement for Aceh Patchouli Oil] (Aceh: Forum Masyarakat Perlindungan Nilam Aceh (FMPNA))
[3] Ma'mun M 2011 Pasca Panen Nilam [Postharvest of Nilam] Bunga Rampai Nilam Status Teknologi Hasil Penelitian Balai Penelitian Tanaman Obat dan Aromatik Ed. Miftahudin E (Bogor: Kementerian Pertanian)
[4] Kusuma H S and Mahfud M 2016 *J. of Applied Res. on Medicinal and Aromatic Plants*
[5] Dennis R M 2004 *Pressure Vessel Design Manual 3rd Edition* (Burlington: Elsivier)
[6] ASME 2010 *ASME Boiler and Pressure Vessel Code An International Code II Part D Properties (Custory) Materials* (New York, USA)

IOP Conf. Series: Materials Science and Engineering **931** (2020) 012011 doi:10.1088/1757-899X/931/1/012011

The Ability of the Fast Fourier Transform to De-Noise a Strain Signal

Teuku Edisah Putra, Husaini, Desy Asrina, and Muhammad Dirhamsyah
Dept. of Mechanical and Industrial Engineering, Faculty of Engineering, Universitas Syiah Kuala, Banda Aceh, Indonesia.

Email: edi@unsyiah.ac.id

Abstract. This study aims to develop an algorithm for filtering the noise in the fast Fourier transform. It requires the determination of the cycles, based on the rain flow counting method. This design was further applied to remove the lower amplitude cycles in a 60-second strain signal. Therefore, the results indicated that the method was able to remove more than 10% of the lower amplitude cycles at a frequency of 45 Hz. The filtered strain signal maintained the fatigue damage by more than 90% and further upheld its original characteristics. This study concluded that the developed algorithm was able to identify and remove the noise contained in the strain signal

1. Introduction

Fatigue failure arises when a material is subjected to repeated static or dynamic stresses. It could occur with a load that is lower than what is presumed to break material with a single impact, below the ultimate tensile strength and the yield strength [1,2]. This phenomenon is very dangerous because it occurs without a deformation, hence it is very difficult to predict since it is influenced by many factors [1]. Researchers constantly try to improve knowledge on experimental prediction, in order to not neglect matters that could precedent to this phenomenon, such as mass, stiffness and damping ratio, collectively known as structural parameters.

Data used to assess fatigue life is obtained by conducting experimental measurements [3]; however, they do not come out pure as they contain noise from the movement of other components, which further disturbs this assessment [4]. Previous research by Abdullah et al. [5] utilized the Daubechies wavelet to conducted noise filtering in a strain signal. However, wavelet is complicated by its process of analysis, starting from the selection of the mother wavelet, the order and the level of operation [6,7]. Since the wavelet transform is difficult to apply generally, this study aims to develop an algorithm for noise filtering with a simpler method based on the fast Fourier transform (FFT).

2. Methodology

Fig. 1 presents the noise filtering algorithm developed using the FFT. This study used a strain signal for validating the noise filtering algorithm. The strain signal was measured on an automotive lower arm, driven on a public road for 60 seconds with the mean value of -11 $\mu\varepsilon$, as shown in Fig. 2. The strain signal gave the fatigue damage of 1,64E-04 damage per block and the number of cycles of 1,412. The material used for the simulation was the SAE 1045 carbon steel, commonly utilized for fabricating the component and its properties are shown in Table 1.

IOP Conf. Series: Materials Science and Engineering **931** (2020) 012011 doi:10.1088/1757-899X/931/1/012011

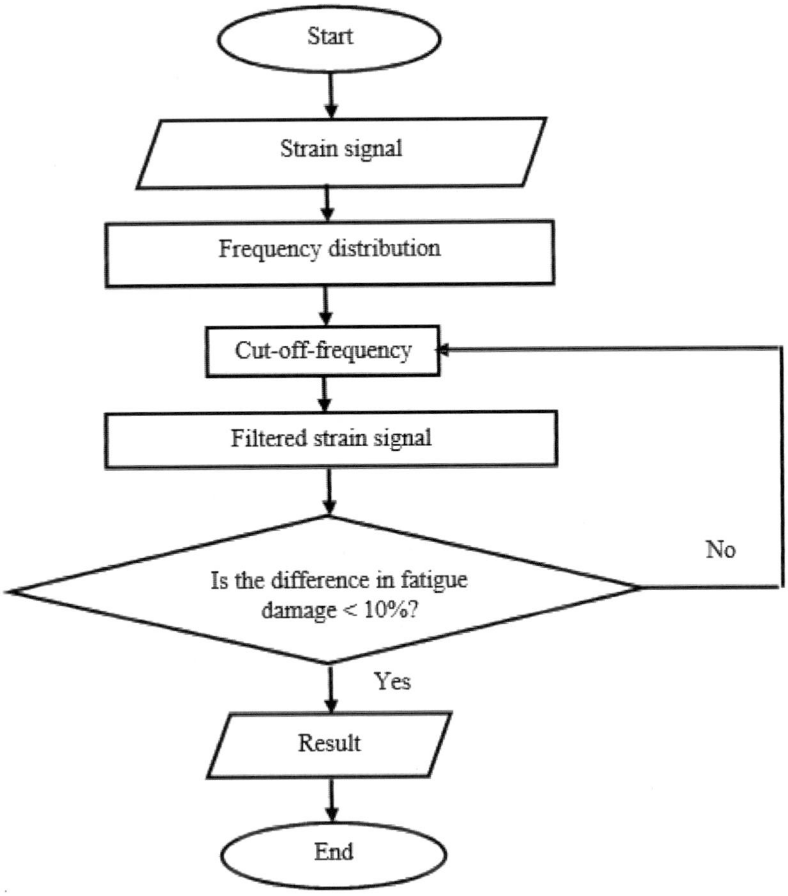

Figure 1. Flowchart of the algorithm

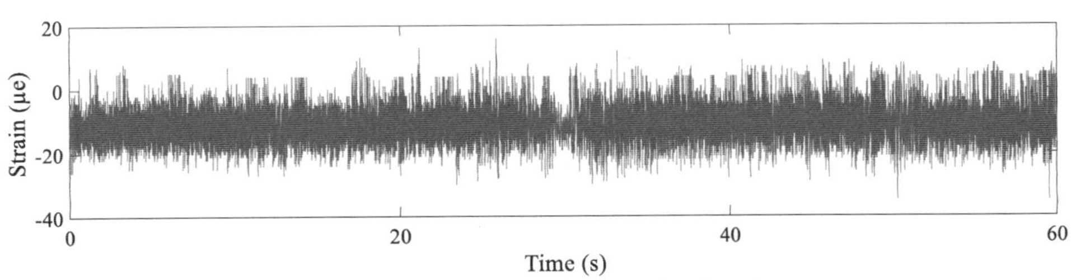

Figure 2. The original strain signal

Table 1. Mechanical properties of the SAE1045 carbon steel [8]

Properties	Values
Ultimate tensile strength, S_u [MPa]	621
Material modulus of elasticy, E [GPa]	204
Yield strength [MPa]	948
Fatigue strength coefficient, σ'_f [MPa]	-0.092
Fatigue strength exponent, b	-0.445
Fatigue ductility exponent, c	0.26
Fatigue ductility coefficient, ε'_f	621

The effectiveness of the algorithm depended on the maintenance of at least 90% of the fatigue damage in the filtered strain signal, in order to keep its original characteristics. Prediction of fatigue life is more accurate with the strain-life approach because it considers the formation of plastic events in the local area. One of which is the Smith-Watson-Topper (SWT) parameter [9], which further gives an accurate result for a signal with positive mean stress [10]. It is defined as:

$$\sigma_{\max}\varepsilon = \frac{\sigma_f'^2}{E}(2N_f)^{2b} + \sigma_f'\varepsilon_f'(2N_f)^{b+c} \tag{1}$$

where σ_{max} is the maximum stress, ε is the strain amplitude, σ'_f is the fatigue strength coefficient, E is the material modulus of elasticity, N_f is the number of cycles to failure for a particular stress range and mean, b is the fatigue strength exponent, ε'_f is the fatigue ductility coefficient, and c is the fatigue ductility exponent.

Rain flow cycle counting method [11] was used to determine the number of cycles, which involved small cycles, treated as interruptions to larger ones. Furthermore, the strains were rearranged from the maximum peak or the minimum valley, starting with whichever was greater in absolute magnitude. Small cycles were extracted first, leaving the larger ones for the end of the process.

Fatigue damage for each loading cycle D_i is represented by:

$$D_i = \frac{1}{N_f} \tag{2}$$

The Palmgren-Miner rule [12,13] was used to determine the cumulative fatigue damage, as in the following:

$$D = \sum\left(\frac{n_i}{N_f}\right) \tag{3}$$

Fatigue damage ranges from zero (indicating no deterioration) to one, which implies an outright failure.

Signal processing is characterized as the recovery of the procedure, based on physically observed data, usually subject to interference caused by differences in thermal, electrical, or mechanical situations. The random characteristics of these signals implore the importance of statistical techniques in the formulation of models in accordance with system behaviour. Furthermore, the process also requires the development of appropriate techniques to estimate model parameters and evaluate performance [14].

Fatigue signals are generally characterized by statistics representing their features. Therefore, a number of statistical parameters were used to monitor patterns and classify random signals. The parameters usually used include the mean, standard deviation (SD), root-mean square (r.m.s.), and kurtosis. For a signal F_j with the number of data n, the mean value $\overline{x}$ is estimated using:

$$\overline{x} = \frac{1}{n}\sum_{j=1}^{n} Fj \tag{4}$$

SD was used to determine how data were distributed in a sample and the number of average variabilities in a data set. SD can be stated as follows:

$$SD = \sqrt{\frac{1}{n}\sum_{j=1}^{n}(F_j - \overline{x})^2} \tag{5}$$

r.m.s. was used to estimate the total amount of energy in a discrete data F_j. This is expressed by:

$$r.m.s = \sqrt{\frac{1}{n}\sum_{j=1}^{n}F_j^{\,2}} \tag{6}$$

Kurtosis is a statistical parameter that is sensitive to spikes in the data with a discrete equation as follows:

$$K = \frac{1}{n(SD)^4}\sum_{j=1}^{n}(F_j - \overline{x})^4 \tag{7}$$

Frequency analysis was performed to convert a time signal into the frequency domain, to be used in obtaining information. It is further distinguished into four types, which include a low and high pass for filtering minimal and enormous frequencies respectively, also, band pass, to filter iterations that pass through, while band rejected screens that get caught [8]. Since noise was generally presented at high frequencies, low pass was used in this study.

The Fourier transform *FT* [15] is defined as:

$$FT_{(\omega)} = \int_{-\infty}^{\infty} F_{j(t)} e^{-i\omega t}\, dt \tag{8}$$

where F_j is the time domain signal, t is the time and ω is the angular frequency, defined by:

$$\omega = 2\pi f \tag{9}$$

where f is the frequency.

According to Figliola & Beasley [16] and Kreyszig [17], the discrete Fourier transform is the main equipment in the frequency domain analysis, where time signals are converted to a set of points. This apparatus is however quite difficult to implement because it involves repeated addition and multiplication operations. Therefore, the FFT was introduced by Cooley & Tukey [18], which effectively breaks a signal down into discrete sinusoidal waves and reduces the repetition required in the digital signal conversion processes.

In addition, the spectrum analysis was performed using the power spectral density (PSD), taking into account the energy of a signal in the frequency domain further showing the strength of the variations in intensity as a function. It is expressed as [19]:

$$PSD = \frac{1}{2\pi}\sum_{n=-\infty}^{\infty} F_{j(n)} e^{-i\omega t} \tag{10}$$

3. Results and Discussion

Based on the frequency distribution presented in Fig. 3, the cut-off frequency (COF) applied was 40 Hz, 45 Hz and 50 Hz since the noise was distributed there. Filtering at 50 Hz was able to remove almost 6% of the cycles and maintained the fatigue damage of 96.5%. At 45 Hz, it maintained 1,268 cycles, ensuring the removal of 10.2% of lower amplitude cycles, which further reduced to 8.4% or changed of the fatigue damage to become 1.5E-4 damage per blok. Clarifying at 45 Hz was selected as the optimum value since the fatigue damage deviation was observed to be more than 10% at 40 Hz, which was detrimental to the original behavior of the strain signal.

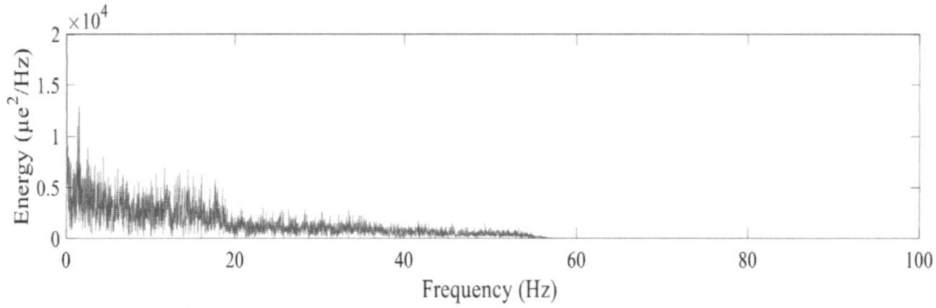

Figure 3. Frequency distribution in the range of 0 – 100 Hz

The 3-D histogram obtained from filtering at 45 Hz shown in Fig. 4 clearly shows that some cycles at the lower ranges (Fig. 4.a) have been removed; hence they did not appear in Fig. 4.b. Therefore, this proved that the developed algorithm was able to identify and remove low amplitude cycles. The removed cycles did not change the statistical parameter values in any way. The original and filtered strain signals provided similar values, which were the mean of -11 με, SD of 2 με, r.m.s. of 12 με and kurtosis of 3.3. Furthermore, the energy maintenance, determined by calculating the area under the PSD [8], was also considered as seen in the plot in Fig. 5 that filtering at 45 Hz reduced the energy by 11%. However, this reduction did not significantly influence the fatigue damage and the original signal characteristics.

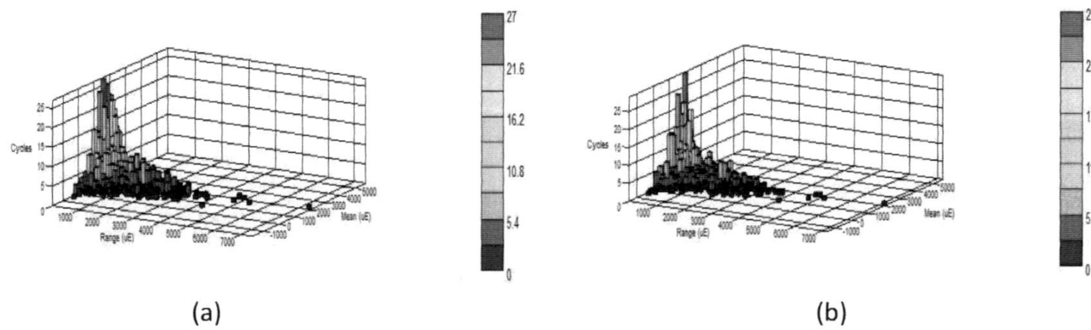

Figure 4. Cycles: (a) original, (b) filtered

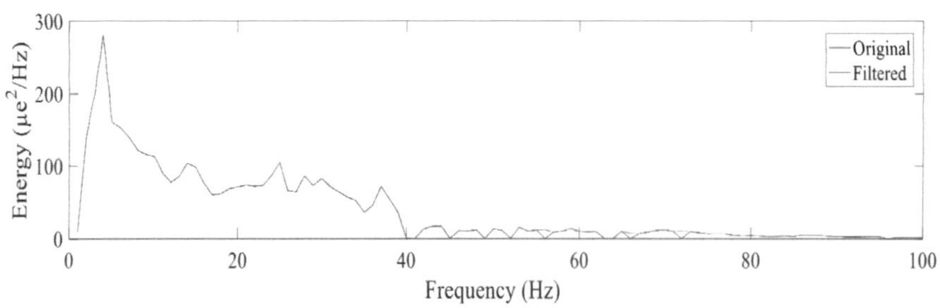

Figure 5. PSD of the original and filtered strain signals

4. Conclusion

An FFT-based algorithm for noise filtering was developed in the current study. Based on the results, this process was able to remove more than 10% of the lower amplitude cycles and maintained more than 90% of the fatigue damage. The cycles was removed up to 44%. This was due to the strain signal, which was a non-stationary, with a kurtosis value of 3.3. A non-stationary signal changes with time, while the time information in the FFT was lost when the analysis was performed, hence it was not relevant to apply this for non-stationary signals [20]. However, the purpose of this study was achieved as the technique did not change the strain signal significantly as removed high frequencies were produced by low amplitude cycles obtaining a minimal effect on the fatigue damage.

Acknowledgements

The authors would like to express their gratitude to Universitas Syiah Kuala for financial support for this research through the grant No. 63/UN11.2/PP/PNBP/SP3/2019.

References

[1] X.W. Ye, Y.H. Su, J.P. Han, *A state-of-the-art review on fatigue life assessment of steel bridges*, Mathematical Problems in Engineering, (2014) 1-13.

[2] K.A. Zakaria, S. Abdullah, M.J. Ghazali, *A review of the loading sequence effects on the fatigue life behaviour of metallic materials*, Journal of Engineering Science and Technology Review, 9 (2016) 189-200.

[3] H.B. Pawar, D.D. Desale, *Optimization of three wheeler front suspension coil spring*, Procedia Manufacturing, 20 (2018) 428-433.

[4] W.Y. Liu. *A review on wind turbine noise mechanism and de-noising techniques*, Renewable Energy, 108 (2017) 311-320.

[5] S. Abdullah., S.N. Sahadan., M.Z. Nuawi, Z.M. Nopiah, *Fatigue road signal denoising process using the 4th order of Daubechies wavelet transform*, Journal of Applied Sciences, 8 (2008) 2496-2509.

[6] B. Tang, W. Liu, T. Song, *Wind turbine fault diagnosis based on Morlet wavelet transformation and Wigner-Ville distribution*, Renewable Energy, 35 (2010) 2862-2866.

[7] Y. Jiang, B. Tang, Y. Qin, W, Liu, *Feature extraction method of wind turbine based on adaptive Morlet wavelet and SVD*, Renewable Energy, 36 (2011) 2146-2153.

[8] nCode, GlyphWorks, nCode International, Ltd., Sheffield, 2018.

[9] K.N. Smith, P. Watson, T.H. Topper, *A stress-strain function for the fatigue of materials*, Journal of Materials JMLSA, 5 (1970) 767-778.

[10] A. Ince, G. Glinka, *A modification of Morrow and Smith-Watson-Topper mean stress correction models*, Fatigue & Fracture of Engineering Materials & Structures, 34 (2011) 854-867.

[11] A. Aid, A. Amrouche, B.B. Bouiadjra, M. Benguediab, G. Mesmacque, *Fatigue life prediction under variable loading based on a new damage model*, Material and Design, 32 (2011) 1883-191.

IOP Conf. Series: Materials Science and Engineering **931** (2020) 012011 doi:10.1088/1757-899X/931/1/012011

[12] A. Palmgren, *Die Lebensdauer von Kugellagern*, Zeitschrift VDI, 68 (1924) 339-341.

[13] M.A. Miner, Cumulative damage in fatigue, *Journal of Applied Mechanics*, 67 (1945) A159-A164.

[14] W. Alexander, C. Williams, Digital Signal Processing - *Principles, Algorithms and System Design*, Elsevier Inc, Amsterdam, 2017.

[15] J. Fourier, *The Analytical Theory of Heat*, The University Press, Cambridge, 1878.

[16] R.S. Figliola, D.E. Beasley, *Theory and Design for Mechanical Measurements*, fifth ed., John Wiley & Sons, Inc., New Jersey, 2011.

[17] E. Kreyszig, *Advanced Engineering Mathematics*, tenth ed., John Wiley & Sons, Inc., New Jersey, 2011.

[18] J.W. Cooley, J.W. Tukey, *An algorithm for the machine calculation of complex Fourier series*. Mathematics of Computation, 19 (1965) 297-301.

[19] J. Wijker, *Random Vibrations in Spacecraft Structures Design: Theory and Applications*, Springer Science+Business Media B.V., New York, 2009.

[20] Y. Blace, C. Hautier, F. Lefebvre, A. Djordjevic, T. Creveaux, I. Rogowski, *Analysis of the tennis racket vibrations during forhand drives: selection of the Mother wavelet*, Journal of Biomechanics, (2017).

IOP Conf. Series: Materials Science and Engineering **931** (2020) 012012 doi:10.1088/1757-899X/931/1/012012

Desorption properties of mechanically milled MgH_2 with double catalysts Ni and SiC

A Rahwanto[1], Z Jalil[1], Akhyar[2], E Handoko[3]
[1]Physics Department, Universitas Syiah Kuala, Darussalam, Banda Aceh, Indonesia
[2]Mechaninal and Industrial Engineering Department, Universitas Syiah Kuala, Darussalam, Banda Aceh, Indonesia
[3]Physics Department, Jakarta State University, Jakarta, Indonesia

E-mail: adirahwanto@unsyiah.ac.id

Abstract. The study of the MgH_2 double-catalyzed Ni and SiC based storage system has been done. The objective of this study is to improve the hydrogen desorption temperature and the hydrogen desorption capacity. The material preparation was done through mechanical alloying technique. In this method, the milling of the material was carried out within 5 hours, 10 hours and 15 hours, with the ratio of the ball to the powder was and the speed of 350 rpm with the catalyst insertion catalyst by 5 wt% and 10 wt% SiC. Results from XRD measurements revealed that the sample was successfully reduced to a nanocrystal scale. The phase emerging from the XRD observation is the phase MgH_2 as the main phase and followed by Ni and SiC phase as the minor phase. From the observation with SEM showing irregular particle shape, the particle size in the sample is not homogeneous because of the agglomeration effect and cold welding that causes the particle size to look large. The hydrogen release capacity was obtained in the samples which were milled for 10 hours at the weight of 5 wt% Ni-SiC catalyst with the release of 2.6 wt%. Based on observation with thermogravimetric analyzer (TGA), the lowest temperature was obtained in the samples which were milled 10 hours with catalyst 5 wt% SiC, which was 365 °C.

1. Introduction

Theoretically, magnesium (Mg) can absorb hydrogen in large quantities (7.6 wt%). This number exceeds the maximum limit targeted by the World Energy Agency of 5 wt% and the absorption/desorption temperature ≤100 °C [1]. Also, Mg is an easy to obtain, affordable light element for research in today's world. But Mg has a drawback, its kinetic reaction is very slow, to absorb the hydrogen takes at least 60 minutes, then its operating temperature is very high (300°C). As to the limits set by the IEA, a material must be able to absorb hydrogen at temperatures below 100°C and in a short time [2-4].

Research on magnesium alloys has been done by several researchers with various methods of manufacture and various alloying elements. Ranjbar et al [6] have conducted a study that SiC particles are evenly dispersed in the MgH_2 matrix, hydrogen absorption/desorption kinetics after inserting a small amount of SiC (5wt%), which can absorb 6.3 wt% hydrogen in a very short time (5 minutes) at 350 ° C.

The study of double catalyzed MgH_2 material on Ni and SiO_2 with SiO_2 3 wt%, 5 wt%, 10 wt%, and 15 wt% catalysts using reactive ball milling technique on thermal testing using DSC showed that the lowest H_2 desorption process among four variations of catalyst that is at 15 wt% happened at temperature 361 ° C [7-8]. From these results, the temperature obtained has decreased. But the temperature is still very high to be applied to the automotive industry, so it needs to do further research using the same material but with a nano-catalyst [9-14]. Very recently, Jalil et al. [15] introduced a double nano-catalyst to improve the characteristics of magnesium hydrides for hydrogen storage material application. In this work, we report the using of dual-catalysts Ni and SiC simultaneously in magnesium hydrides (MgH_2) prepared by high energy planetary ball milling.

2. Materials and method

The materials used were MgH_2 99.99% in purity, 50 μm in size, both from Sigma Aldrich. Then, Ni (with 99% purity and 50 nm particle size) and silicon carbide (SiC), both purchased from Hongwu Nano. The three ingredients are mixed and then divided into two parts, namely MgH_2 + Ni + 5 wt% SiC; and MgH_2 + Ni + 10 wt% SiC are milling with time variation for 5 hours, 10 hours and 15 hours. The sphere and material ratio is 10: 1 with a sample weight of 1 g at 350 rpm for each sample. A small sample is transferred for characterization. Samples were characterized by X-ray diffraction (XRD) (Cu-Kα radiation with λ = 1,54060 Å) to determine the phase that appeared. The morphology of the samples was obtained by using SEM (JEOL JSM-5310LV) and the thermal properties of hydrogen desorption were observed using Thermogravimetric Analyzer (TGA).

3. Results and discussion

The identification on MgH_2 + Ni + SiC sample using X-Ray Diffractometer (Shimadzu D6000) from sample with catalyst variation of SiC (5 wt% and 10 wt%) and have been milled for 5 hours, 10 hours and 15 hours, produces X-ray diffraction patterns (Fig. 1 and 2). Then, to know the phase contained in the sample, the comparison of the experimental data with the standard data using the database of files issued by the Joint Committee for Powder Diffraction Standard (JCPDS) has been done.

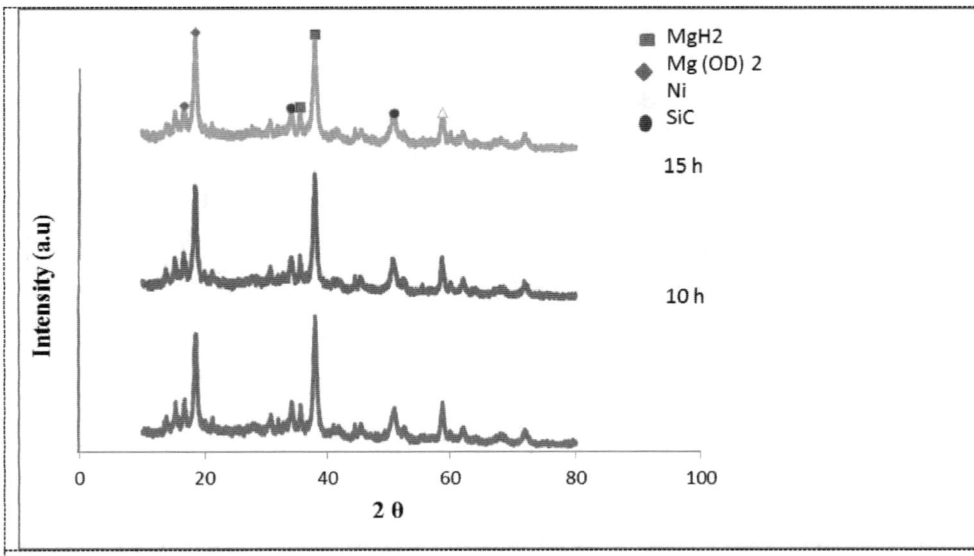

Figure 1. XRD profile on MgH_2 + Ni + SiC material with catalyst variation 5 wt% SiC.

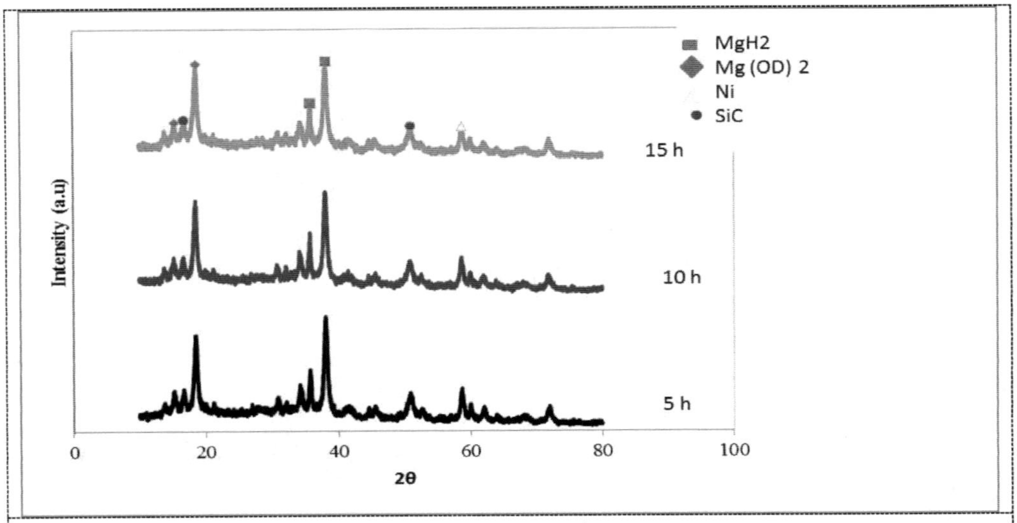

Figure 2. XRD profile on MgH_2 + Ni + SiC material with catalyst variation 10 wt% SiC.

In the diffraction pattern shown, it is clear that the MgH_2 phase appears as the main phase at an angle of 37.9°. The MgH_2 phase is also seen appearing at an angle of 2θ = 16.6°, the Ni phase appearing at an angle of 2θ = 35.6, and followed by the SiC phase that appears at peaks of 2θ = 58.6°, and 31.1°. However, during the weighing and removal process from the sample site to the milling container, the material is contaminated with air so that the emergence of $Mg(OD)_2$ at an angle of 2θ = 50.7° and 18.4° as an impurities that may affect the process of hydrogen desorption temperature, likewise seen in Figure 2 it is clear that the phase MgH_2 appears as the main phase at an angle of 37.7 °. The MgH_2 phase is also seen appearing at an angle of 2θ = 16.6 °, the Ni phase appearing at an angle of 2θ = 35.6, and followed by the SiC phase that appears at peaks of 2θ = 58.6 °, and 31.1 °. However, during the weighing and removal process from the sample site to the milling container, the material is contaminated with air so that the emergence of $Mg(OD)_2$ phase at an angle of 2θ = 50.7 ° and 18.8 °.

Ni and SiC in this study worked as a catalyst and were used only in very small amounts. The catalyst serves to speed up the reaction but does not react, as shown in Figure 1. The emergence of Ni and SiC phases is caused because Ni and SiC do not participate in react in MgH_2, this is indicated by no change of phase of Ni and SiC. The emergence of $Mg(OD)_2$ phase in the XRD test results is due to the process of weighing the sample and moving from the sample to the milling container, the material is contaminated with the outside air causing the oxygen to enter the material.

The peak widening distribution in Figure 1 shown from the FWHM value in each sample did not show any significant change, meaning that the difference of FWHM value in each sample is very small, causing the diffraction pattern to look the same. In the sample with 5 wt% Ni catalyst insertion with SiC (5 and 10 wt%) catalyst variations and 5, 10 and 15 hour milling time.

The FWHM value shows the peak width of the diffraction pattern. From the XRD diffraction pattern, it can be concluded that the increasing catalyst wt% SiC FWHM value and the resulting smaller crystal size means the diffraction peak and the crystalline size decreases with increasing catalyst from 5wt% to 10 wt% SiC and inversely proportional to milling time, milling time is larger the FWHM value and grain size. The widening of the peaks indicates that the MgH_2 phase crystallinity has occurred. In the previous study, the decrease in crystal size can decrease the temperature of onset and temperature. Smaller crystal sizes will form larger surface areas when compared to larger crystal sizes. This results in hydrogen making recombination easier.

Ranjbar et al [6] mentioned that in addition to the reactive milling process, the role of silicon carbide (SiC) can accelerate grain refinement. Because SiC has the hard properties which are generally carbide metals. With a fine grain size, the hydrogen will more easily enter the hydrogen atomic grid and the more hydrogen capacity absorbed. Furthermore, as reported in [6] that the milling process in the atmosphere of hydrogen, either with pure Mg or MgH_2 has been able to reduce the size of the ductile Mg grains. One more thing, it is believed that the hardness of the carbide element in SiC is also capable of reducing the grain size to cause the MgH_2 surface area to become larger. With the widening of this surface area will ultimately have a positive effect on the desorption properties of MgH_2.

The result of surface morphology with SEM obtained information that the dimension of sampling grains with time 10 hours is smaller than the dimension of sampling grain with 5 hours. The results as shown in Figure 3 and 4.

Figure 3. The SEM image of the sample MgH_2 + Ni + 5 wt% SiC after 5 hours ball milling.

Figure 4. The SEM image of the sample MgH_2 + Ni + 5 wt% SiC after 10 hours ball milling.

However, as previously reported [6] the milling process is believed that the hardness of the Ni element can reduce grain size to cause the MgH_2 surface area to become larger. With the widening of this surface area will ultimately have a positive effect on the properties of MgH_2 absorption and desorption. However, the number of Ni catalysts in MgH_2 should be limited, because if more Ni doping will cause negative effects such as low absorbed hydrogen capacity, slow reaction kinetics and can also block the diffusion process of H_2.

Gravimetric testing is required to see the hydrogen release capacity of the double catalyzed-MgH_2 sample. In Fig. 4. showed the desorption characteristic of MgH_2 + Ni + 5wt% SiC with 10 hours of the milling process. It was shown that MgH_2 + Ni + 5wt% SiC material with 10 hours of milling process was able to hydrogen as much as 1.2 wt% in 16 minutes at the T onset 365 °C. It can be explained that smaller particle sizes will result in a large contact surface area between the samples and the hydrogen, resulting in the hydrogen chamber entering the sample particles larger and consequently the hydrogen absorption rate increases [14]. Despite the reaction relatively fast (in 16 minutes), the hydrogen release is still high (desorption/T onset at 365 °C). Dai et al. [19] have studied that the dehydrogenation improvement of Ni doped MgH_2 system is expectable due to the formation of thermodynamically less stable Mg_2NiH_4 phase. It was found that desorption property is improved by surface distortion & formation of second phase.

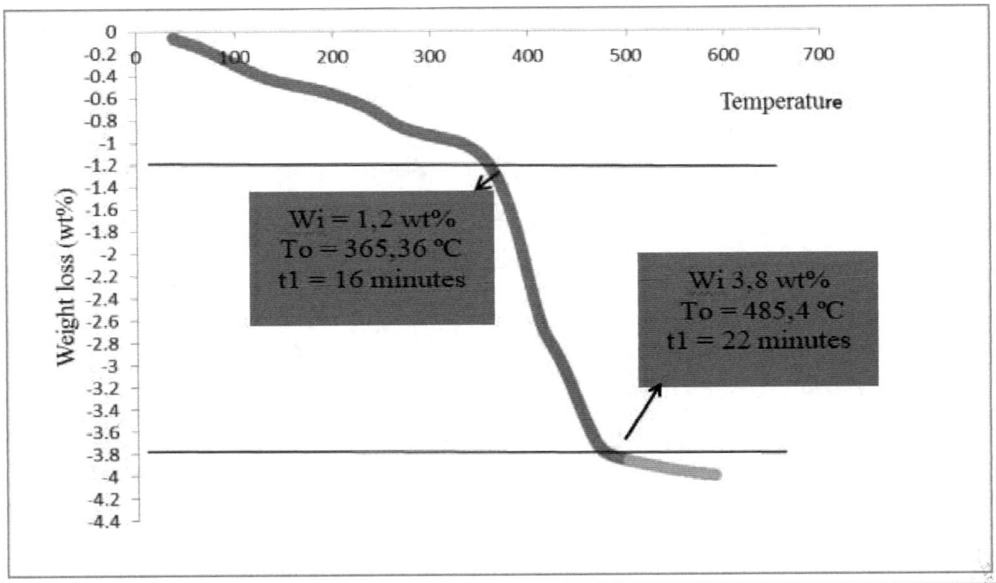

Figure 5. TGA profile of MgH_2 + Ni + 5wt% SiC milled for 10 hours.

Based on observation with TGA, the lowest temperature was obtained in the samples which were milled 10 hours with catalyst 5 wt% SiC, which was 365 °C. Goudarziafshar et al. [16] has studied that the effect of Ni atom doped on SiC nanocage on the molecular adsorption of hydrogen for practical hydrogen storage. It is important that the substituted Ni atom is capable of adsorbing two hydrogen molecules at room temperature. The effect of nanostructure catalyst was also studied as one of the important role in the improvement of magnesium hydrides as solid hydrogen storage [17]. El-Khatabi et al [18] have investigated that double substitution will effect to the reduction of heat of desorption and preservate the optimum gravimetric capacity. It was also reported that the co-doping have desorption temperatures within the optimum range and therefore suitable for the practical applications.

4. Conclusion

The study on the use of double catalysts Ni and SiC on the desorption of MgH_2 for solid hydrogen storage material with variation of wt% SiC has been successfully done. It can be concluded that MgH_2 catalyzed with Ni and wt% SiC variations in small amounts (5 wt% and 10 wt%) synthesized by high energy planetary ball milling technique of 350 rpm was successful, and the desorption properties of MgH_2 material can be improved. SEM analysis shows that Ni and SiC can act well as a catalyst. TGA profile inform that the desorption temperature was obtained at 365 °C for 5 wt% SiC. However, despite the fast reaction of double catalyzed on MgH_2 kinetic, the desorption temperature is still high for the application.

5. Acknowledgement

The authors want to express their gratitude to Syarifah Fathmiah, S.Si and Karmila Sari, S.Si, M.Si (Materials Physics Lab., Unsyiah) responsible for XRD and thermogravimetric analyzer test.

6. References

[1] Schlapbach L and Zuettel A 2001 Hydrogen storage materials for mobile applications *Nature Vol 414*

[2] Sellier D A, Aymoniera E and Bobet J L 2009 Hydrogen sorption properties of magnesium particles decorated with metallic nanoparticles as catalyst *Journal of Alloys and Compounds 476* **152-159**

[3] Barkhordarian G, Klassen T and Bormann R 2006 Catalytic Mechanism of Transition Metal Compounds on Mg Hydrogen Sorption Reaction *J. Phys. Chem. B, 110* **11020 – 11024**

[4] Song M Y, Kwon S N, Bae J S and Hong S H 2009 Hydrogen-storage properties of melt spun Mg-23.5 wt%Ni milled with nano Nb_2O_5 *Journal of Alloys and Compounds 478* **501-506**

[5] Oelerich W, Klassen T and Bormann R 2001 Mg-based hydrogen storage materials with improved hydrogen sorption *Mater. Trans. 42* **1588-1592**

[6] Ranjbar A, Guo Z P, Yu X B, Wexler D, Calka A, Kim C J and Liu H K 2009 Hydrogen storage properties of MgH_2–SiC composites *Mater Chem Phys 114* **168–172**

[7] Sun L and Gong K 2001 Silicon-Based Materials from Rice Husks and Their Applications *Ind. Eng. Chem. Res. 40* **5861-5877**

[8] Akihiro Y, Makoto S and Junichi K 1998 Mechanical alloying and solid phase reaction of Mg-SiO_2 system *Funtai Funmatsu Yakin Kyokai Koen Gaiyoshu* **85**

[9] Jalil Z, Rahwanto A, Mulana F and Mustanir Desorption temperature characteristic of Mg-based hydrides catalyzed by nano-SiO_2 prepared by high energy ball milling *International Journal of Technology 7* **1301-1306**

[10] Jalil Z, Rahwanto A, Handoko E and Mustanir 2017 The role of nano-Ni catalyst in MgH_2 obtained by reactive mechanical milling method for solid hydrogen storage application *AIP Conference Proceedings 1826* **020002**

[11] Jalil Z, Rahwanto A, Handoko E and Mustanir 2017 Magnetic behavior of natural magnetite (Fe_3O_4) extracted from beach sand obtained by mechanical alloying method *AIP Conference Proceedings 1862* **030023**

[12] Larsson P, Araujo C M, Larsson J A, Jena P and Ahuja R 2008 Role of catalysts in dehydrogenation of MgH_2 nanoclusters *PNAS 105* **8227–8231**

[13] Jalil Z, Rahwanto A, Sofyan H, Usman M and Handoko E 2018 The use of Silica from beach sand as catalyst in Magnesium based hydrides for Hydrogen storage materials *IOP Conf. Series: Earth and Environmental Science 105* **012093**

[14] Jalil Z, Rahwanto A, Akhyar H, Razali R and Handoko E 2018 MgH_2-SiC based hydrogen storage material prepared by reactive mechanical alloying method *IOP Conf. Series: Earth and Environmental Science 105* **012098**

[15] Jalil Z, Rahwanto A, Ismail I, Sofyan H and Handoko E 2018 The use of nano-silicon carbide and nickel as catalyst in magnesium hydrides (MgH_2) for hydrogen storage material application, *Mater. Res. Express 5* **064002**

[16] Goudarziafshar H, Abdolmaleki M, Moosavi-zare A R, and Soleymanabadi H 2018 Hydrogen storage by Ni-doped silicon carbide nanocage: A theoretical study *Physica E: Low dimensional Systems and Nanostructures 101* **78-84**

[17] Jalil Z, Rahwanto A and Handoko E 2018 Effect of nanostructure Ni on the sorption properties of mechanical milled MgH_2 *MATEC Web of Conferences 197* **02014**

[18] Khatabi M E, Naji S, Bhihi M, Benyoussef A, Kenz A E and Loulidi M 2018 Effects of double substitution on MgH_2 hydrogen storage properties: An Ab initio study *Journal of Alloys and Compounds 743* **666-671**

[19] Dai J H, Song Y and Yang R 2011 Intrinsic mechanisms on enhancement of hydrogen desorption from MgH_2 by (001) surface doping *International Journal of Hydrogen Energy 36* **12939-12949**

IOP Conf. Series: Materials Science and Engineering **931** (2020) 012013 doi:10.1088/1757-899X/931/1/012013

Body Segment Dimensions of Indonesian Male Army

M. Iqbal[1], R. Salam[2], I. Hasanuddin[1] and A. B. Hassan[3]

[1] Department of Mechanical and Industrial Engineering, Faculty of Engineering, Universitas Syiah Kuala, Banda Aceh, Indonesia

[2] Aceh Regional Police, Banda Aceh, Indonesia

[3] Electrical, Electronic and Automation Section, Universiti Kuala Lumpur Malaysian Spanish Institute (UniKL – MSI), Kulim High Technology Park, 09000, Kulim, Kedah, Malaysia.

E-mail: mohd.iqbal@unsyiah.ac.id

Abstract. The paper discusses the size of six body segments of the Indonesian Male Army, i. e. stature, leg length, chest depth, chest width, spine length and shoulder high. The particular body segments were believed as the most important body part that related to the strength and load handling capability of humans as reported by researchers. Two hundred and sixty-eight (268) male soldiers have participated as the subjects of this research. The size of the body segments and the body weight were measured by using measuring tape, Martin type anthropometer and digital body weighing scale. The result was analysed to observe the significance of the data as well as to determine the mean, standard deviation, 5th percentile and 95th percentile of the data. The comparison of the mean stature with Indonesian male students and male army of other countries was discussed in the paper. In future research, the result of this study will be used to predict the load handling capability of the Indonesian male army using the biomechanics method.

1. Introduction

Matching of supply, job task, work procedure, tool, and equipment to anthropometry has been an important measure in ergonomic. Ergonomic consideration has taken an important place in the design of tasks and facilities in industries in order to fix the nature of the human body in term of size, posture [1], load [2] and movement [3]. The continuous improvement could be easily found in many types of hand tools that become more comfortable to users over the years. The ergonomic improvement could be found in the end product as well, such as vehicles, computers and furniture. The user could enjoy a better driver seat of their cars compared to the same product of the previous years. As a result, better ergonomic design leads to a better health level of users, fewer defects of the product or service, and higher productivity.

The size of body segment, body area and body weight have been studied for decades for the purpose of safety and comfort of humans. Body size was used as a reference for the design of the products such as hand tools, machines, buildings, vehicles, furniture, clothes, etc. [4, 5, 6]. It has been a challenge for the manufacturer since it becomes a consideration of the costumer's decision. The manufacturer should improve the ergonomics of their products in order to win the market competition.

The body weight was used as a reference for the load received by the products such as chairs, vehicles, shoes, etc. The body weight of users will be very much considered during the design stage of the product, i.e. in material selection and mechanical design development. With the help of computer design software, even the dynamic load of the product could be predicted by using the body weight data of users.

Research of anthropometry was needed to be conducted to all human races since different races have different body sizes [7, 8, 9]. That was the reason why all countries should support anthropometry research on their people. The data could be used to fit the people of that particular race to their daily used products. As humans grow, the different body sizes would be found in a different group of ages.

Hartono [10] reported that the stature of the human body increased by the increase of the age from 6-9 years old to 10-12 years old. In the older age, the stature did not increase, but the volume and the weight of the body increased significantly. However, the elderly subjects lost some volume of their body size, as well as their body weight. As the food quality and the wealth of living improves, the anthropometry of population changes after some period of time. Tomkinson et al. [11] reported that there were substantial changes (typically increases) in the body dimensions of male Australian Army personnel over the period 1977-2012. Therefore the anthropometry data need to be updated periodically.

Some groups of the profession require special body sizes, such as fashion models, flight attendance, pilot and soldiers. Fashion model and flight attendance require high stature and slim body size for beauty and attractive appearance, while soldiers require special body stature and size for strength, stamina and capability to carry the load. Bauer and Freivalds [12] and Brackley and Stevenson [13] studied the load-carrying capability by using a biomechanics approach based on anthropometry measure. They discussed six body segments that significantly affected the load-carrying capability, i.e. stature, leg length, chest depth, chest width, spine length, and shoulder height. The workers should not carry the load over their capability because it could cause injuries, generate mental stress and burnout syndrome and decrease the working performance, as reported by Erwan et al. [14] and Hasanuddin et al. [15].

This study aims to investigate the anthropometry of the Indonesian army. The research focuses on the six body segments as investigated earlier [12, 13]. The result could be used as a reference to predict the strength, stamina and load-carrying capability, as well as the size of military products, i. e. clothes, tools and weapons. In addition, the result of the investigation was compared to the anthropometry data of Indonesian university students to identify the influence of occupation in anthropometry data. We also compared the anthropometry data of the Indonesian male army with the male army of other countries to learn the influence of race in anthropometry data of military persons.

2. Methodology

There were 40 body segments commonly considered as anthropometric body measurement as shown in Fig.1. Each of the body segments was measured either during the standing or sitting position. However, this study did not focus on the anthropometry measurement of the army. The study observes the size of body segment that reflect the performance of the army in term of strength and capability to handle the load.

Two hundred and sixty-eight Indonesian male army soldiers (mean age = 27.6 years, SD = 6.93 years) from Kodam Iskandar Muda (Indonesian Army in Aceh States Unit) participated in the study. The soldiers were recruited from the Infantry Battalion 112/ Raider Banda Aceh. No soldiers from other districts were involved in the study. Even though anthropometrics difference existed among different ethnicities in Indonesia [7], the subjects of this study did not represent only a particular ethnic since the Indonesia Military Management distributes the soldiers proportionally to all state units in Indonesia. In other words, the subjects of the study reflected national soldiers rather than Aceh (ethnic) soldiers only.

The survey was conducted at the indoor activity building of Infantry Battalion 112/ Raider Banda Aceh. The size of the body segment and the body weight were measured by using a measuring tape, Martin type anthropometer, and digital body weighing scale. The measurement and data recording process were taken by research assistants that have been well trained by senior and experienced researchers. The topics of the training include the measurement procedure, how to handle the equipment, how to deal with the subject's confidentiality, how to handle the complaint, and how to record the data. The measurement process of body segment was shown in Figure 2.

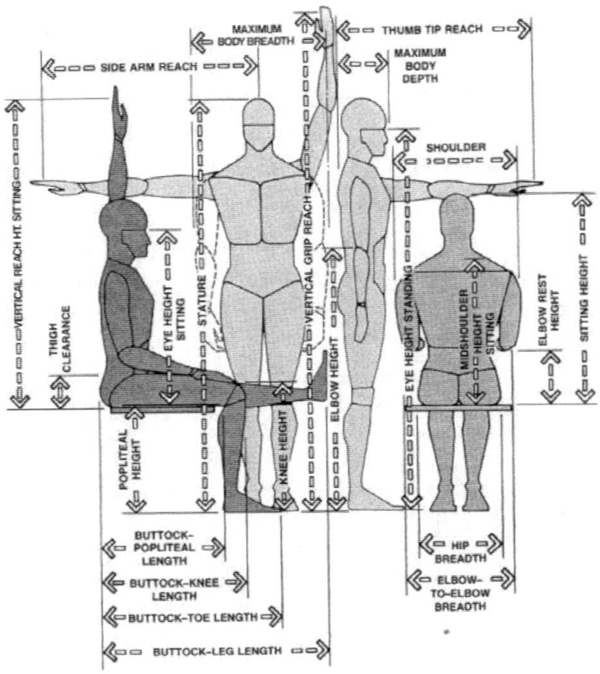

Figure 1. Anthropometric body measurement

During the exercise and real duty, a soldier carries a backpack with a standard weight and size. The soldier must be able to carry the backpack during a long journey and within a long duration of time. Most studies found that an acceptable weight limit of a subject should be between 10% and 15% of their body weight [12, 13]. However, an acceptable weight limit did not only depend on body weight but the size of body segments as well.

A study conducted by Kroemer [16] showed that individual differences such as body dimensions should be considered to determine the safe limit for the load to be carried. The study was followed by Ismaila and Owaba [17] that used the spinal shrinkage principles to determine the maximum weight of a load that a worker should lift to be safe.

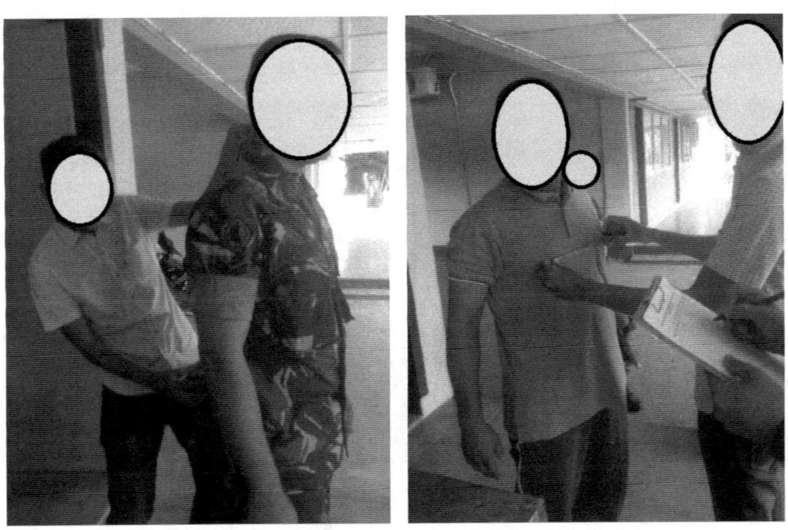

Figure 2. The measurement of body segments

IOP Conf. Series: Materials Science and Engineering **931** (2020) 012013 doi:10.1088/1757-899X/931/1/012013

Further study by Ismaila [18] used a strain energy-based model in terms of the size of body segments, Young Modulus of Elasticity of articular cartilage and maximum permissible spinal shrinkage to predict the capability of the subject to carry the load in the backpack during both standing and moving. The study aimed to determine the safe backpack mass limit based on the consideration of the subject's anthropometric characteristics.

Based on the above literature and methods, only six body segments were observed in this study. The particular body segments have been proven as the most important body part to represent the strength and load handling capability of the subjects [18,19]. They were stature, leg length, chest depth, chest width, spine length, and shoulder hight. All of the six body segments were measured during a standing position, as shown in Fig 2.

3. Result and Discussions

The sizes of the six body segments of all 268 subjects were recorded and further analyzed. The mean and standard deviation (SD) of each body segment dimension were calculated using a simple average and sample standard deviation equations [20]. The value of the 5th percentile and 95th percentile were calculated based on a normal statistical distribution. The result of the data collection and calculation are shown in Table 1.

As shown in Table 1, the average stature of the Indonesian male army was 1690.9 mm, with 90 percent of stature between 1650 mm and 1760 mm and the SD of 39.5 mm or 2.45% of the average stature. Comparison between the stature of the Indonesian male army and the stature of Indonesian male university students shows different values. As reported by [8, 10], the average stature of Indonesian male students was 1670.1 mm, and its SD was 53.1 mm or 3.18% of its average stature. It means that the Indonesian male army was taller than Indonesian male students. On the other hand, the deviation (the difference between the shortest and the highest) of the Indonesian male army's stature was smaller compared to the deviation of Indonesian male students' stature.

Sutalaksana and Widyanti [21] reported the influence of occupational when comparing the anthropometry data of Indonesian students and traditional Indonesian workers. They concluded that the students come from better socioeconomic status compared to the traditional workers, and this is associated with better nutrition, better child care, and better medical and social services, which result in bigger anthropometry size. However, the factor contributing to the differences above does not apply to the difference between the Indonesian male army and Indonesian male students. The difference was because of a minimum stature and a narrow gap limit of minimum and maximum stature that must be fulfilled by army candidates when they apply to be a trainee at Indonesian Military School. Whereas, there are no criteria of body stature for any candidate of university students.

Table 1. Body segment dimensions of male Indonesian army, all units are in mm

No	Body segments, weight, and age of the subjects	Mean	Standard Deviation	5th Percentile	95th Percentile
1	Stature	1690.9	39.5	1650	1760
2	Leg length	1027.1	117.4	940	1450
3	Chest depth	224.0	36.1	140	300
4	Chest width	231.6	31.5	190	320
5	Spine length	523.8	45.8	470	620
6	Shoulder high	1454.7	44.2	1380	1520
7	Body weight (kg)	70.28	7.59	58	86
8	Age (Year)	27.6	6.93	20	42

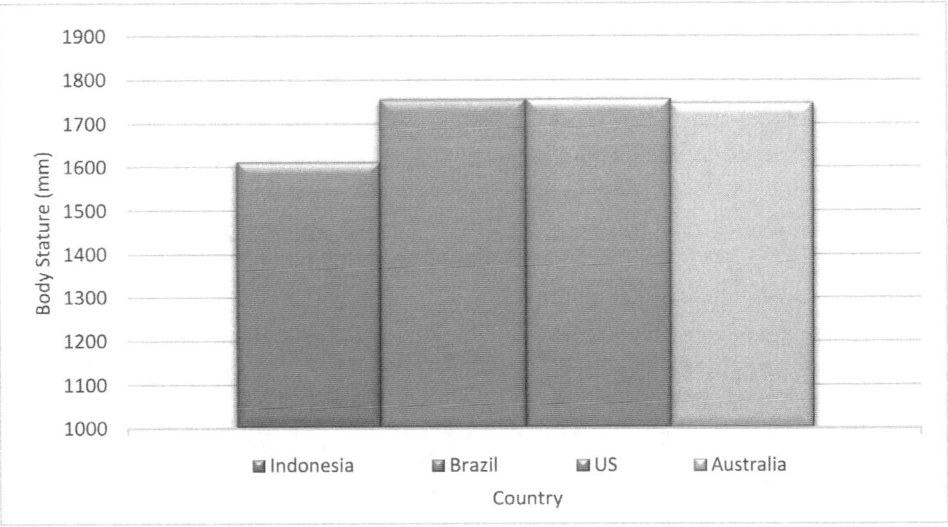

Figure 3. The comparison of army's body stature (mm) among 4 countries

The comparison between the stature of the Indonesia army and three other countries (Brazil, United States (US) and Australia) is shown in Figure 3. The stature of the army of Brazil, US and Australia was 1755, 1756 and 1747 mm respectively [11]. Fig 3 shows that the Indonesian army was shorter than the army from the other three countries. The US army is the tallest, where the Brazilian is in the second place.

Tomkinson et al. [11] also reported that the body size of the Australian army had been changed in time. There has been a significant change in 12 anthropometry measures of the Australian army between 1977 and 2010-2012. Actually, time increases in body size have been reported for people of all ages over the past couple of centuries. It is believed that the increases in body size to be due to nutrition and access to health care, improvements in environmental conditions, leading to the reduction of growth-inhibiting factors.

The body stature of the Australian army in 1977 was 1747 mm. It was 38 mm shorter than their stature in 2010-2012. However, it was still 56 mm higher than the Indonesian army of the present year (2018) discussed in this paper. There is no reference found that could be used to compare the situation of Brazil, US, and Australian army to the Indonesian army. Further research is required to find out whether the difference in body size among the military reflects their strength and capability.

The body stature did not really represent the performance of a soldier. A taller soldier has a heavier body weight, so they require more metabolic energy during the activity such as walking, running and handling the load. Usually, the metabolic energy and stamina of a subject are not significantly linear with their body weight and stature [4]. Furthermore, the army of Indonesia, Brazil, US and Australia came from different races. It has been proven that different races have different anthropometry size [8,10]. However, the anthropometry data of the army from Indonesia's neighbour countries, such as Malaysia, Thailand, Singapore, the Philippines and Brunei Darussalam, are not available in the literature. Therefore, the comparison to the same race cannot be properly discussed in this paper.

4. Conclusions

The result of the observation of six body segments of the Indonesian male army has been presented and discussed in the paper. The comparison of the average stature between the Indonesian male army and Indonesian male students showed that the Indonesian male army was taller than Indonesian male students. However, the Indonesian male army has a narrower deviation of stature. Comparison of the stature of the Indonesian male army to the other three countries' male army showed that the US male army was the tallest, followed by Brazil, Australia, and Indonesia. The result of the study can be used in future study to predict the load handling capability of the Indonesian male army using the biomechanics method.

Acknowledgments

The authors would like to acknowledge the management of Kodam Iskandar Muda (Indonesian Army in Aceh States Unit) and the Infantry Battalion 112/ Raider Banda Aceh. Special appreciation is extended to all soldiers that have participated as the subject of this survey.

References

[1] R. Z. R. Umar, C. F. Ling, N. Abdullasim, N. Ahmad, I. Halim, M. Hamid, *Occupational wrist postural assessment and monitoring system: development and initial validation*. Journal of Engineering Science and Technology 14 (6) (2019) 3421-3426.

[2] M. Iqbal, M. Tadjuddin, I. Hasanuddin, *Mathematical Models to Determine the Metabolic Energy Consumption of Stoop and Squat Lifting*. Proceedings of International Conference on Natural and Environmental Science (ICONES) (2014) 61-66

[3] H. Soewardi, A. A. Rahmayani, *Analysis of Muscle Contraction on Pottery Manufacturing Process Using Electromyography (EMG)*. IOP Conf. Series: Materials Science and Engineering 105 (2016) 012023.

[4] S. Pheasant, C. M. Haslegrave, Bodyspace - A*nthropometry, Ergonomics and The Design of Work (3rd ed.)*, CRC Press, Taylor & Francis Group, United States of America. (2006).

[5] J. Chang, K. Jung, J. Hwang, Y. Kang, S. Lee, A. Freivalds, *Determination of bicycle handle diameters considering hand anthropometric data and user satisfaction*, Proc Hum Factors Ergon Soc Annu Meet. 54(20) (2010) 1790-1793.

[6] D. Lacko, T. Huysmans, J. Vleugels, G. De Bruyne, M. M. Hulle, J. Sijbers, S. Verwulgen, *Product sizing with 3D anthropometry and -medoids clustering*, Computer-Aided Design. 91 (2017) 60-74.

[7] B. Yue, K. M. Varadarajan, S. Ai, T. Tang, G. Li, *Differences of knee anthropometry between chinese and white men and women*, The Journal of Arthroplasty. 26(1) (2011) 124-130.

[8] A. Widyanti, L. Susanti, I. Sutalaksana, K. Muslim. *Ethnic differences in Indonesian anthropometry data: evidence from three different large ethnics*, International Journal of Industrial Ergonomics. 47 (2015) 72-78.

[9] W. Kim., Y. M. Kim, M. H. Yun, *Estimation of stature from hand and foot dimensions in a Korean population*, Journal of Forensic and Legal Medicine. 55 (2018) 87-92.

[10] M. Hartono, "*Indonesian anthropometry update for special populations incorporating drillis and contini revisited*, International Journal of Industrial Ergonomics. 64 (2018) 89-101.

[11] G. R. Tomkinson, N. Daniell, A. Fulton, A. Furnell, *Time changes in the body dimensions of male Australian Army personnel between 1977 and 2012*, Applied Ergonomics. 58 (2012) 18-24.

[12] D. H. Bauer, A. Freivalds, *Backpack load limit recommendation for middle school students based on physiological and psychophysical measurements*, Work. 32 (2009) 339–350.

[13] H. M. Brackley, J. M. Stevenson, *Are children's backpack weight limits enough? A critical review of the relevant literature*, Spine. 29(19) (2004) 2184–2190.

[14] F. Erwan, M. Iqbal and I. Hasanuddin, *The analysis of nursing's work systems in relation to burnout syndrome (A case study: nurses in RSUDZA, BLUD RSIA, and RSUD Meuraxa*, Banda Aceh, Indonesia), IOP Conference Series: Materials Science and Engineering. 453 (2018) 012051.

[15] I. Hasanuddin, M. Iqbal, Suparno, R. A. R. Ghazilla, F. Erwan, *The effect of working environment and motivation towards labors' performance: a case study in Aceh at P. T. Pupuk Iskandar Muda*, Aceh International Journal of Science, 8(1) (2019) 35-40.

[16] K. H. E. Kroemer. *Ergonomic Design of Material Handling Systems*. Lewis Publishers, New York, Buca Raton. (1997).

[17] S. O. Ismaila, O. E. A. C. Owaba, *Quantitative approach for establishing safe weight of lift*, Engineering Review. 32(1) (2012) 1–8.

[18] S. O. Ismaila, *Safe backpack weight limit for secondary school students in Ibadan, Southwestern Nigeria*. Alexandria Engineering Journal, 57 (2018) 547-554.

[19] A. J. F. Paula, J. C. V. Silva, J. C. R. P. Silva, *The influence of load imposed by the backpack school in children and teens in Brazil*, Procedia Manufacturing. 3 (2015) 5350–5357.

[20] M. Kouchi, K. Mochimaru, Tsuzuki, T. Yokoi, *Random errors in anthropometry*, J. Hum. Ergol. 25 (1996) 155-166.

[21] I. Z. Sutalaksana, A. Widyanti, *Anthropometry approach in workplace redesign in Indonesian Sundanese roof tile industries*, International Journal of Industrial Ergonomics. 53 (2016) 299-305.

IOP Conf. Series: Materials Science and Engineering **931** (2020) 012014 doi:10.1088/1757-899X/931/1/012014

Economic study in simulating 5 MWp solar farm planning with 2 technologies under Indonesia Feed-in Tariff in Weh Island-Aceh, Indonesia

S Rahmawati[1], **M Iqbal**[2], **I D Sara**[3]
Industrial Engineering Department, Universitas Syiah Kuala, Darussalam, Banda Aceh, Indonesia
Department of Mechanical and Industrial Engineering, Universitas Syiah Kuala, Darussalam, Banda Aceh
Department of Electrical Engineering, Universitas Syiah Kuala, Darussalam, Banda Aceh

E-mail: mohd.iqbal@unsyiah.ac.id

Abstract. Up to now in Weh Island-Aceh, the power energy used comes from the Diesel Power Plant (PLTD) only, where the cost of this electricity production (BPP) reached more than Rp 2,500/kWh. This price is currently higher than the requirement set by the Minister of Energy and Mineral Resources of Rp 1,733/kWh. Because of these significant differences, special attention is needed to examine the natural resources that can be converted into electrical energy with the aim that the energy offered can reduce the price of electricity production. The purpose of this research is to conduct a study that analyses the economic feasibility of investment by considering all economic aspects using HOMER Software. The result shows that the overall value of an investment with Indonesia's feed-in tariff is feasible. The cost of energy production with si-mono and si-poly technologies was Rp 1,051/kWh and Rp 914/kWh respectively. By the capacity of 5000 kWp, return on investment (ROI) can occur in 10.73 years with si-mono technology and in 8.87 using si-poly technology from 25 years of the current system period. Using the Bank's interest rate of 5.25%, the Internal Rate of Return (IRR) is considered feasible to apply because these two technologies illustrated the greater number than the general bank interest rate prevailing in Indonesia.

1. Introduction

Currently, the primary source of electrical energy used in the world comes from fossils, which are known to be environmentally unfriendly and are limited in capacity. Electricity demand continues to increase every year; therefore, the energy crisis becomes one of the significant issues of concern in the world. Up to 2016, Aceh had a peak load of 325 megawatts (MW) derived from fossils such as gas power plants (PLTGs) and some from steam power plants (PLTU), which are distributed from the Sumatra region. The availability of electricity in this region should be able to reach 50% of the peak load requirement, or there must be around 500 MW, yet currently, there are only 340 MW available. This condition causes the importance of additional electricity supply; one alternative is to propose environmentally friendly and renewable energy. In addition, it was also carried out to support the achievement of the ASEAN target to increase the energy mix of renewable energy (EBT) by 23% in the coming year 2025, where the world's EBT has supplied 24.5% of the total energy needs consisting of 16.6% hydro, 4% wind, 2% biopower and 1.5% solar PV [1]. Also Considering that Indonesia's potential with two seasons is very supportive of this EBT procurement, one of which is a solar plant with the utilization of solar radiation, where the light obtained from the sun can then be converted into electrical energy using PV panels.

In fact, the investment in EBT implementation, especially in Indonesia, is still costly [2]. The amount of EBT investment costs will significantly affect investors' interest to invest, whereas, on the other hand, Indonesia's target is severe enough to be able to develop EBT. Therefore, one of the reasons for the Indonesian government to support the achievement of this EBT target is to impose a Feed in Tariff (FiT) rule as a benchmark for the purchase price of EBT by the State Electricity Company (PLN). The regulation stipulated by the Minister of Energy and Mineral Resources (ESDM) No. 17 of 2013 [3] is that the FiT for the Aceh region is \$ 17 cents and can reach up to \$ 30 cents if using products manufactured or assembled domestically. The next most recent regulation is No.12 in 2017 [3] where FiT is set at 85% of the electricity production cost (BPP) in the local area.

The following are projections of electricity demand by sector and electricity production until the year 2050 in Indonesia:

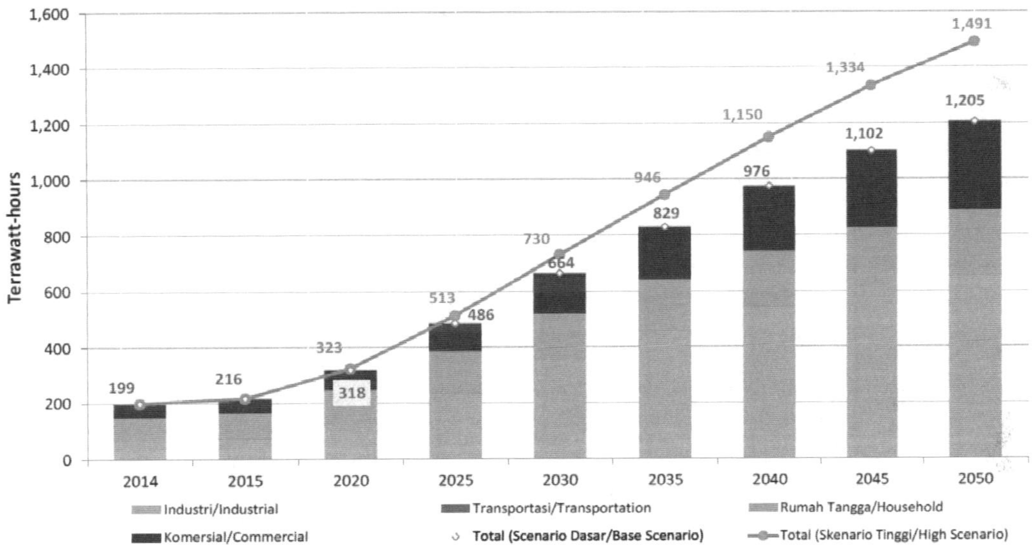

Figure 1. Projections for electricity demand by sector and electricity production by 2050 in Indonesia (Source: Indonesia's energy outlook 2016, BPPT)

The economic analysis on this project, especially solar farm, aims at producing information needed as a reference for investors to determine the feasibility of an investment today and in the future [4,5,6,7]. In this study, the planning will be simulated based on the real electricity peak of Weh Island. The fact that the use of EBT, especially solar energy farming on Weh Island, has been developed four years ago, some obstacles make this electrical energy plan fail and be left out. Such obstacles include the condition of the grass that grows tall, lack of skilled technicians, and the lack of socialization about the maintenance of the PV system, so that the impact makes PLTS inoperable [8].

IOP Conf. Series: Materials Science and Engineering **931** (2020) 012014 doi:10.1088/1757-899X/931/1/012014

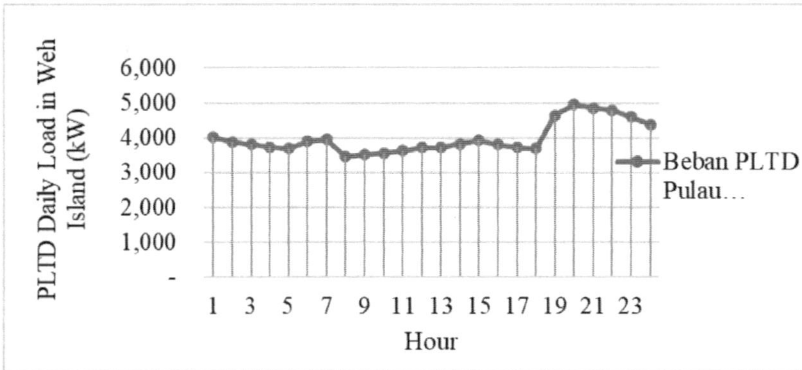

Figure 2. Daily load curve of PLTD Weh Island for the past year (April 2017-March 2018) [8]

2. Methodology

In this study, all data were taken in Weh Island located at the coordinate point of 5.84° N 95.28° East, where average radiation is 5.1 kwh/m2/day, simulated in 2 technologies of panels, si-Monocrystalline, and si-Polycrystalline. Designed for 5 MWp (grid-connected system) based on the peak of electricity daily load in this location. Information for this economic analysis, management planning, and price of all materials needs (e.g., solar PV module (si-mono and si-poly), inverter, land acquisition) was gathered from a consultant solar PV company (PT. Rekasurya Prima Daya). The FiT rates use of 85% from the real cost of electricity from diesel power (set by the government).

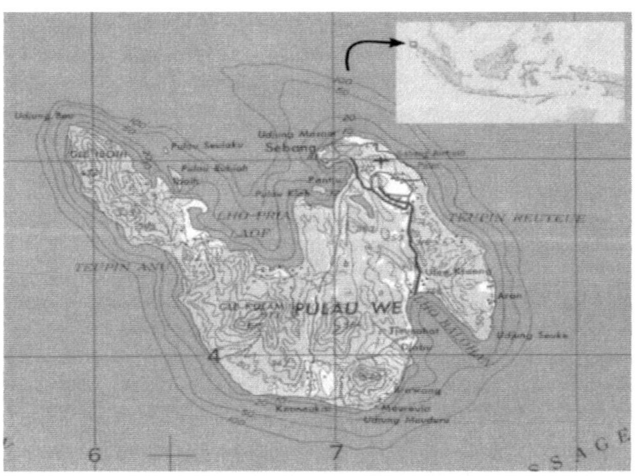

Figure 3. Map of Weh-Aceh Island Location for Solar Farm Planning

2.1. HOMER (Hybrid Optimization of Multiple Electric Renewables)

HOMER software was developed by The National Renewable Energy Laboratory (NREL), USA in collaboration with Mistaya Engineering, where copyright is protected by the Midwest Research Institute (MRI) and used by the United States Department of Energy (DOE). HOMER is used to design hybrid power generation systems by combining conventional energy and renewable energy [9]. Homer Energy software used to analyse the Simulation, optimization of the systems offered along with the sensitivity analysis of all Renewable Energy (EBT) systems. The three main objectives HOMER can do are simulation, optimization and sensitivity analysis. HOMER is designed specifically to analyse and optimize the renewable energy industry system, as implemented by [4], [10]. HOMER is widely used to study, measure and analyse the performance of grid-connected PV systems.

2.2. Data simulation

2.2.1 Solar radiation
Data by the National Aeronautics and Space Administration (NASA) [11] shows the average annual radiation on this island is 5.1 kWh / m2 / day. The highest solar radiation usually occurs in February to April about 5.75 kWh / m2 / day, while the lowest radiation occurred in October and November of 4.66 - 4.67 kWh / m2 / day.

Table 1. Data on the potential of solar radiation in Weh Island - Aceh (kWh/m2/day)

Solar radiation in Weh Island	Jan	Feb	Mar	Apr	May	Jun	Jul	Aug	Sep	Oct	Nov	Dec	Avrg
Monthly radiation (kWh/m2/month)	167.1	161	179.5	169.2	155.3	149.7	151.9	149.1	142.2	144.8	139.8	151.3	155.1
Daily radiation (kWh/m2/day)	5.39	3.75	5.79	5.64	5.01	4.99	4.9	4.81	4.74	4.67	4.66	4.88	5.1

2.2.2. Sizing of solar farm

2.2.2.1. Using PVsyst for Sizing All Capacity Needs for 5 MWp Solar Farm
Optimization of material required for 5 MWp has been calculated by PVSYST software in the previous study by Rahmawati et al., 2019, [12]. This 5MWp solar farm planning is based on the geographical location of Weh Island-Indonesia, which is at latitude 5.830 North - longitude 95.320 East with a time zone of UT + 7, the altitude of 95 m and Albedo at the point of 0.20. Meteom data were obtained from NASA-SESE satellites from 1983-2003. The slope design of the panel was set at an angle of 100 eastward and the Azimut 00 provision with fixed tilt, and the model used in the simulation is the Perez transposition by ignoring the Shading factor.

2.2.2.2. HOMER simulation

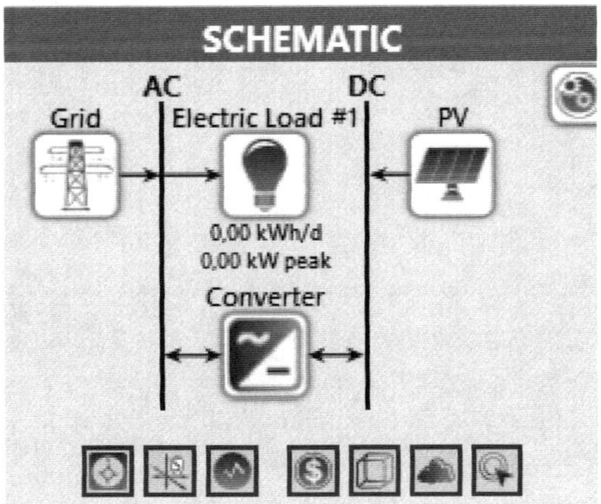

Figure 4. Input for HOMER model simulation

2.2.2.3. PV Generation
The number of materials required for 5 MWp solar farm was 16,668 panels with utilizing 300 Wp [12] with 25 years of life time and different specification. Si-mono technology with the efficiency of 16.4% costs Rp. 134.062.550.000 while si-poly technology with an efficiency of 15% costs Rp. 118.022.795.000

IOP Conf. Series: Materials Science and Engineering **931** (2020) 012014 doi:10.1088/1757-899X/931/1/012014

2.2.2.4. Inverter model

The inverter used in this study is a grid-connected inverter type. The capacity of the inverter was adjusted to the capacity of available power, i.e. 5,000 kWp. The inverter has a capacity of 20 k with Maximum Efficiency > 98.2% [12]. The cost of this inverter was Rp. 14,961,375,000.

2.2.2.5. Economic analysis

This study [13] states that there are several factors that can help investors assess the feasibility of an investment, as follows:

Life cycle analysis (LCC); LCC refers to the sum of all costs, both recurrent and non-recurring with respect to products, structures, systems, or services, during their lifetime. LCC calculation is as follows [4]:

$$LCC = C_{capital} + C_{O\&M} + C_{replacement} - C_{salvage} \quad (1)$$

The capital cost ($C_{capital}$) is the overall price of expenditure to manage the whole system. Capital costs include all prices of components, installation and project design. The costs for operational and maintenance ($C_{O\&M}$) include salaries for workers, maintenance costs for each solar farm product and system monitoring during the system operation. Replacement cost ($C_{replacement}$) is the cost to replace the product at a certain period of the lifetime product, for instance, the replacement of a converter where the lifetime of this product is 15 years. Salvage cost is the residual value of an item or the residual value of an investment. An item that has expired can still be resold. In the calculation of HOMER software, it is estimated that the salvage or residual value of this investment is 20% of the initial investment.

Table 2. Detail cost for 5 MWp solar power generations (si-Mono technology)

No	Description	Weh Island (Rp)	Percetage
1	Management Cost	Rp 3,400,000,000	2.3%
2	Land Acquisition	Rp 10,500,000,000	7.0%
3	Building, Infra and Civil Work	Rp 21,064,000,000	14.1%
4	Solar Module	Rp 75,000,000,000	50.3%
5	Inverters	Rp 14,961,375,000	10.0%
6	Installation Cost	Rp 14,000,000,000	9.4%
7	Electrical Equipment and Asset	Rp 8,092,550,000	5.4%
8	Others (Legal Fees/ General)	Rp 2,000,000,000	1.3%
	Total	Rp 149,017,925,000	100.0%

Table 3. Detail cost for 5 MWp solar power generations (si-poly technology)

No	Description	Weh Island (Rp)	Percentage
1	Management Cost	Rp 3,400,000,000	2.6%
2	Land Acquisition	Rp 10,500,000,000	7.9%
3	Building, Infra and Civil Work	Rp 21,064,000,000	15.8%
4	Solar Module	Rp 59,166,666,667	44.4%
5	Inverters	Rp 14,961,375,000	11.2%
6	Installation Cost	Rp 14,000,000,000	10.5%
7	Electrical Equipment and Asset	Rp 8,092,550,000	6.1%
8	Others (Legal Fees/ General)	Rp 2,000,000,000	1.5%
	Total	Rp 133,184,591,667	100.0%

Net Present Value (NPV); The NPV shows a lump-sum with a certain discounted current which gives a figure of how much the current business value (Rp) is. In this method, the NPV includes a time value factor, to consider all project cash flows, measuring absolute amounts, so that it is easy to follow its contribution to efforts to increase company revenue [13]. To calculate the NPV, we need data about estimated investment costs, operating costs, and maintenance, as well as estimated benefits from the planned project, NPV is explained in the equation as follows [14].

$$\mathrm{NPV} = \frac{\mathbf{TAC}}{\mathbf{CRF}} \tag{2}$$

Where, TAC- total annual cost includes the sum of the annual costs of each system component (Rp), CRF - capital recovery factor (%).

After these costs were calculated, the unit of CoE produced by the generating electricity system could be calculated. The cost of energy (in Rp) is the ratio between the total costs per year of the system with the energy it produces during the same time period [4]. To calculate the price for the energy generated by the system, the following formula was used:

$$\mathrm{CoE} = \frac{\mathbf{LCC}}{\sum_{1}^{n} \mathbf{Epv}} \tag{3}$$

Levelized Cost of Energy (*L*CoE); Levelized Cost of Energy is the price at which electrical energy is generated from certain energy sources and stops functioning even during a certain period of time [13]. LCOE can also be considered the average minimum cost of electricity that can be offered during the duration of the project. LCoE calculations are shown in the following equation:

$$\mathrm{LCoE} \quad = \frac{1+\sum_{t=1}^{n}\frac{LCC}{(1+r)^t}}{\sum_{t=1}^{n}\frac{Et}{(1+r)^t}} \tag{4}$$

Pay Back Period (PBP); Payback period is a parameter that calculates how fast an investment is needed to be returned, so the units used are years, months and days [13].

$$\text{Payback period} = \frac{\textbf{Initial Investment}}{\textbf{Cash In Flow Per Period}} \tag{5}$$

Internal Rate of Return (IRR); Internal Rate of Return (IRR) is a parameter used to obtain an interest rate that equates the total present value of the expected cash flow recipient with the total present value required for investment [15]. The IRR decision criteria are, if the IRR value is greater than the general interest rate in effect, the project is accepted, but if it is the opposite, the project will be rejected. The general interest rate applicable in Indonesia is based on the benchmark interest of all banks, namely, Bank Indonesia, i.e. 5.25% [16]. The formula used is provided in the following [13]:

$$\mathrm{IRR} = \mathrm{i} + \left(\frac{\mathbf{NPV}}{\mathbf{NPV^1 - NPV^2}}\ (\mathrm{i}^1 - \mathrm{i}^2)\right) \tag{6}$$

Where (i) is the interest rate, i^1 is interest rates that produce a positive NPV value, i2 is Interest rates that produce a positive NPV, NPV1 is a positive of net present value, and NPV2 is a negative of net present value.

3. Result and discussion

3.1. Technical analysis

The simulation was conducted after calculating the sizes of all components for 5 MWp, and the use of the components is illustrated in Table 5. The output of the electricity produced by the system with these 2 technologies, as illustrated in Figure 9, gave differences in electricity output, although the technologies used have different efficiency.

Table 4. Results of simulation sizing 5 MWp solar farm with PVsyst software [12]

Sizing Simulation	5000 MW (300 Wp @panel)
Number of Modul	16.668
Number of inverters (20kW)	250
Configuration	18 modules in series with 926 string
The area of land required by the panel	32.309 m2

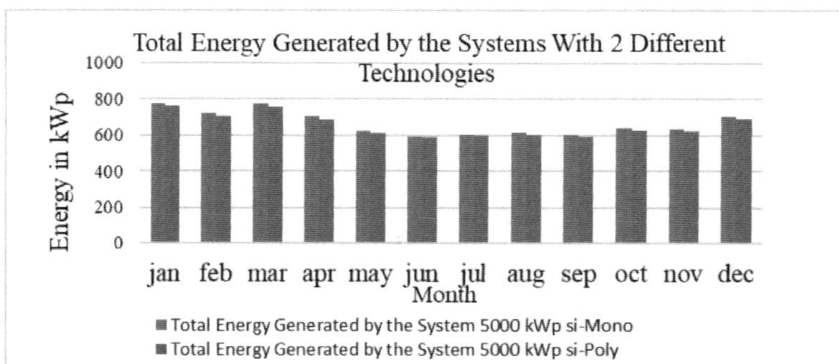

Figure 5. Total energy generated by the systems with 2 different technology

5000 kWp si-Monocrystalline technology

With an investment of Rp. 149,023,925,000, this project is calculated to have an NPV (Net Present Cost) value of Rp. 171,741,079,862. The costs to be incurred annually for the system are Rp. 8,347,640,000, where Energy of as much as 7,991,000 kWh is generated per year with a project life of 25 years. The minimum price that can be offered (LCoE) from Homer simulations is Rp. 1,035, which shows the profit from sales per kWh, and thus the minimum selling price that can be offered is Rp. 1,065 / kWh. The IRR interest rate where NPV = 0 will be obtained at 9.8%, which is greater than the bank interest rate of 5.25%; thus, it is considered feasible. Return on Investment (ROI) generated is 6.4%. Simple payback period, i.e. cash flow with profit without considering the value of money, will occur in year 9.63. Return on capital will occur in year 10.73.

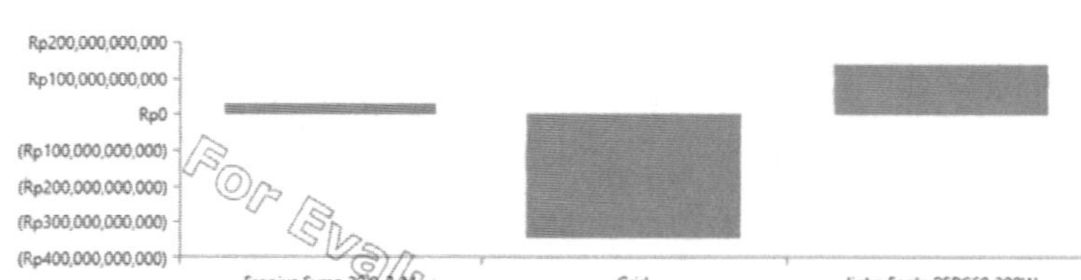

nponent	Capital (Rp)	Replacement (Rp)	O&M (Rp)	Fu	Salvage (Rp)	Total (Rp)
Fronius Symo 20.0-3-M	Rp14,961,375,000.00	Rp23,984,599,071.92	Rp0.00	R	(Rp5,150,041,895.65)	Rp33,795,932,176
Grid	Rp0.00	Rp0.00	(Rp345,513,887,320	R	Rp0.00	(Rp345,513,887,320
Jinko Eagle PERC60 300W	Rp134,062,550,000.00	Rp0.00	Rp5,914,325,281.57	R	Rp0.00	Rp139,976,875,281
System	Rp149,023,925,000.00	Rp23,984,599,071.92	(Rp339,599,562,038	R	(Rp5,150,041,895.65)	(Rp171,741,079,862

Figure 6. HOMER result of cost summary for 5 MWp si-mono technology

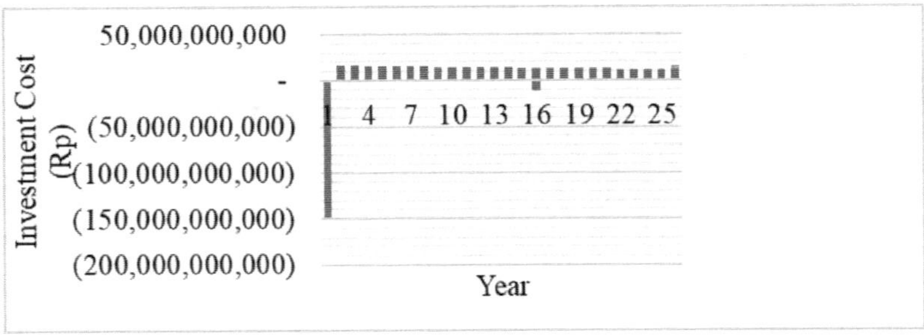

Figure 7. Life cycle cost of 5000 kWp si-monocrystalline technology

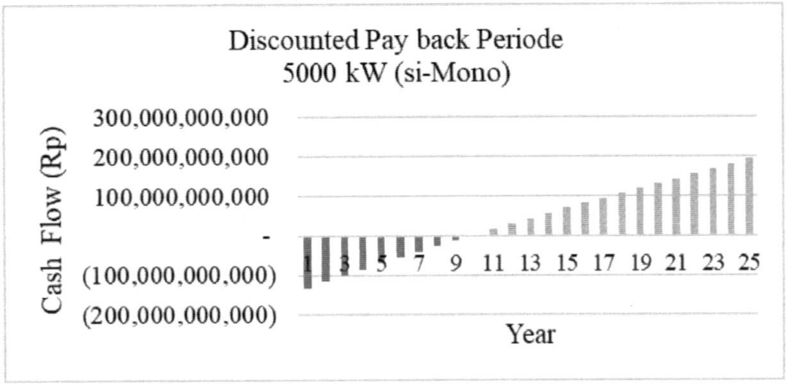

Figure 8. The total cost of the system for 25 years

5000 kWp si-polycrystalline technology

With an investment cost of Rp.133,189,325,000, this project has an NPV (Net Present Cost) value of Rp. 192,021,200,400, costs to be incurred per year for the system are Rp. 9,314,790,000. The energy that rose annually is 7,852,000 kWh with a project life of 25 years. With a sales tariff of 85% from BPH Weh Island, the results show that the minimum price that can be offered (Levelized cost of energy) from Homer's simulation shows an LCOE figure of Rp. 1,177. The minimum selling price that can be offered to the public is Rp. 923 / kWh. The IRR interest rate where NPV = 0 will be obtained is at 11%, which is greater than the bank interest requirement of 5.25% so that it is considered feasible. Return on Investment (ROI) generated is 7.9%. A simple payback period, i.e. cash flow with profit without considering the value of money, will occur in year 8.32. Return on capital will occur in year 8.85. More information can be seen in Table 6.

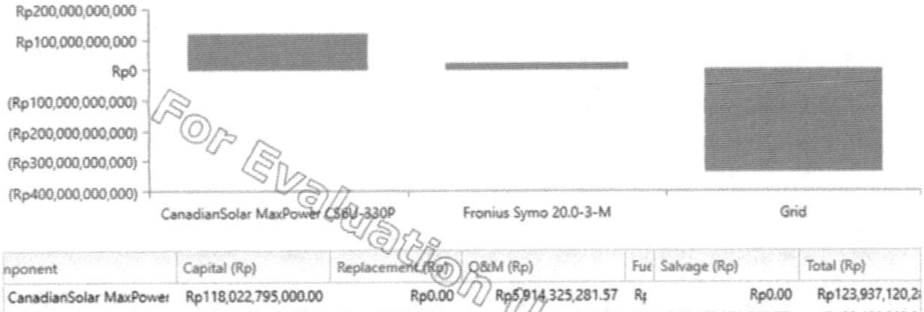

nponent	Capital (Rp)	Replacement (Rp)	O&M (Rp)	Fue	Salvage (Rp)	Total (Rp)
CanadianSolar MaxPower	Rp118,022,795,000.00	Rp0.00	Rp5,914,325,281.57	Rp	Rp0.00	Rp123,937,120,2
Fronius Symo 20.0-3-M	Rp14,961,375,000.00	Rp11,958,950,217.	Rp0.00	Rp	(Rp3,433,361,263.77)	Rp23,486,963,9
Grid	Rp0.00	Rp0.00	(Rp339,445,284,637.21)	Rp	Rp0.00	(Rp339,445,284,63
System	Rp132,984,170,000.00	Rp11,958,950,217.	(Rp333,530,959,355.64)	Rp	(Rp3,433,361,263.77)	(Rp192,021,200,40

Figure 9. HOMER result of cost summary for 5 MWp si-mono technology

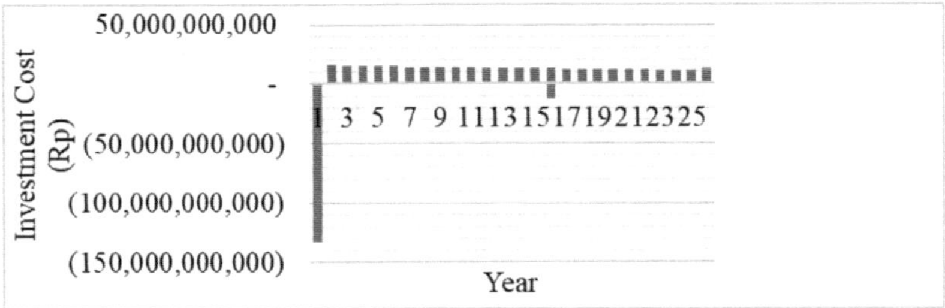

Figure 10. Life cycle cost of 5000 kWp si-polycrystalline technology

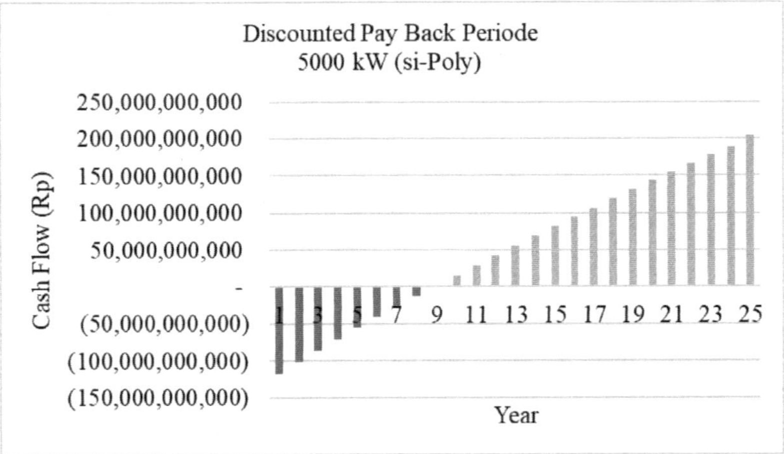

Figure 11. The total cost of the system runs for 25 years

Table 5. The result of economic analysis for 5000 kWp solar farm

Decision Variable	Standart	5 MWp Mono	5 MWp Poly
E_output (kWh/year)		7991	7852
Total Investment (Rp)		149,023,925,000	133,189,325,000
Fit (Rp)		2100	2100
CoE (Rp)		1,035	1,177
LCoE (Rp)	900	1,065	923
NPV (Rp)		172,866,400,000	193,146,600,000
Annual Cost (Rp)		8,347,640,000	9,326,958,000
ROI (%)		6.4	7.9
IRR (%)	5,25%	9.8	11
simple PBP (year)	25	9.63	8.32
Discounted PBP (year)	25	10.73	8.85

4. Summary

In this economic study, several important factors that determine investment decisions have been analysed. This is to provide more detailed information on business actors. The two technologies used were the feasibility information in this project. However, several factors contributed to the decision. Based on return on investment, with FiT sales of Rp. 2,100, -. Although solar farm with si-mono technology produces higher electricity output, cheaper of CoE and greater LCoE, the fastest return on investment was with a solar farm system with 5000 kWp capacity of si-poly technology. Overall, capital recovery in this system occurred in 8.32 with an IRR of 11%.

5. Reference

[1] n d 2016 *Renewable Energy Outlook for Asean Renewable Energy Market Analysis Southeast Asia* International Renewable Energy Agency (IRENA) Asean Centre of Energy (ACE), Tech

[2] Bakri 2016 Aceh Butuh Tambahan Energi Listrik [Aceh Needs Additional Electrical Power] Serambi Indonesia Available [online] htpp://Serambinews.com Accesed on 20 oktober 2017

[3] Ministerial Decree of ESDM No.17 year 2013 and No. 12 year 2017

[4] Pauzi M Kassim M Karam Al-Obaidi M Arkam M Munaaim C and Salleh A M 2015 Feasibility Study on Solar Plant Utility Grid under Malaysia Feed-In Tariff *American Journal Of Engineering and Applied Sciences* **8** 2

[5] Salasovich M J and Gail 2011 *Feasibility Study of Economics and Performance of Solar Photovoltaics at the Refuse Hideaway Landfill in Middleton Wisconsin* (National Laboratory of The U.S. Department of Energy Office of Energy (NREL))

[6] Rangkuti S G and Ch Ramadhan 2016 Perencanaan Pembangkit Listrik Tenaga Surya di Atap Gedung Harry [Solar Power Plan at Harry's Rooftop Building] *Jurnal Rekasurya Elektrika* **9**

[7] Aman S N M and Hadiyono A 2015 Analisis Keekonomian Penerapan Pembangkit Listrik Tenaga Surya Pada Sistem Ketenagalistrikan Nias [Economic Analysis on the Application of Solar Power on Electrical Power System in Nias] *Jurnal Ketenagalistrikan dan Energi Terbarukan* **14** 2

[8] n d 2017 Beban Listrik di Aceh [Electrical Load in Aceh] PT. PLN Pusat Aceh and PLTD Rayon Pulau Weh Sabang Aceh A Report 2405

[9] Index Homer [Online] Available at Homerenergy.com accessed on 20 july 2018

[10] Ikhsan R Sara I D and Lubis R S 2017 Studi Kasus Kelayakan Penerapan Sistem Hybrid Building Applied Photovoltaics (BAPV)-PLN pada Atap Gedung Politeknik Aceh [A Feasibility Study on the Implementation of Hybrid Building Applied Photovoltaics (BAPV)-PLN Systems on Aceh Polytechnic Rooftop] *Jurnal Rekayasa Elektrika* **13** 1

[11] Index PVsyst [Online] Avalaible at http://www.pvsyst.com accessed on 20 July

[12] Rahmawati S Iqbal M Sara I D 2019 Energy Analysis Of Solar Farm Planning In Weh Island, Aceh, Indonesia *International Conference on Design, Energy, Materials and Manufacture*

[13] Sugirianta I B K Giriantari I A D Kumara I NS 2016 Analisa Keekonomian Tarif Listrik Pembangkit Listrik Tenaga Surya 1 MWp bangle dengan Metode Life Cycle Cost [Economic Analysis of Solar Power Tariff 1 MWp bangle using Life Cycle Cost] *Jurnal Teknologi Elektro* **15** 2

[14] Soeharto and Imam 2001 *Studi Kelayakan Proyek Industri* [A Feasibility Study of Industrial Project] 1 Ed (Jakarta, Indonesia: Erlangga)

[15] n d 2013 *The Purchasing Electric Power by PT Perusahaan Listrik Negara (Persero) from Photovoltaic Solar Power Plants, Regulation of the Energy and Mineral Minister* Resources No.17

[16] BI Rate [Online] Available at https://www.bi.go.id/ accessed on 20 July 2018

IOP Conf. Series: Materials Science and Engineering **931** (2020) 012016 doi:10.1088/1757-899X/931/1/012016

A Review on Current Mill Liner Design and Performance

Husni Usman[1], Syarizal Fonna[2] and Sulaiman Thalib[2]
[1]Mining Engineering, Faculty of Engineering, Universitas Syiah Kuala, Banda Aceh, Indonesia
[2]Dept. of Mechanical and Industrial Engineering, Faculty of Engineering, Universitas Syiah Kuala, Banda Aceh, Indonesia

E-mail: husmanphd@unsyiah.ac.id

Abstract. Milling management in Mining and Mineral Industries should keep on improving liner design based on operating experience, ore types, good design practice, and also additional data from ball trajectory and DEM simulations. The small-scale trials for evaluating different liner shapes would rather provide important basic information to industrial mills. The lifters with the same face angle, but different configurations could provide improvements in energy efficiency and performance. Integrating the wear rates of liners from the actual measurement of the changes in liner profile over milling time and also using laser-based thickness gauging technology would assist in establishing a more efficient liner profile.

1. Introduction

As semi-autogenous (SAG) mills become increasingly larger in size, the design of SAG mill liners has been more and more important, and the appropriate liners would enhance mill efficiency and performance. The replaceable liners are mounted inside the mill and exposed to media and/or the material being ground. The liners have to resist impact, be wear-resistant, and provide the most favorable motion of the charge. The primary purposes of the liners are to transfer energy to the charge (ore/rock + grinding media), to hold grinding media, and to protect the mill shell from the wear, as well as to prevent liners/lifters and balls from damage.

The early sets of liners that are normally selected at the milling-plant design stage almost never perform very well in tumbling mills and only limited number of liners work optimally in new installation [1]. Thus, in order to determine proper liner design, it is necessary for the milling management to keep improving the design by relying on useful information from practical operating experience.

Both the liner design and selection of liner materials are particularly crucial in the design and operating stages. They are mainly affected by the physical characteristics of ores (such as abrasiveness, hardness), corrosion environment, and size of balls. The other factors, such as mill speed and filling, also have significant impacts. From a practical experience point of view, the design usually plays the most important role in affecting the mill efficiency and liner life.

The discrete element method (DEM) has provided significant contributions to the design and operation of grinding mills [2]. Because DEM capability to take account of changes in lifter geometry, ball/rock size distribution, and density in predicting mill power draft is better than well-established mill power models, DEM has successfully been used for predicting SAG and ball mill power in large-scale operations. DEM has become increasingly interesting as a tool in establishing the basis for shell lifter designs. However, the leap from the DEM simulation to plant practice is big, particularly for predicting mill power draw with respect to lifter shapes and rock/ore variability. Therefore, the DEM utilized for optimizing the lifter shapes should be validated with small-scale trials rather than plant trials in which the risk, cost and time appear enormous.

The small-scale mill would allow for better evaluating the effect of changes in lifter design, grinding media, and operating parameters, which could then provide important basic information to the design and operation engineers in industrial mills [3,4]. The optimal mill liner is expected to improve the mill efficiency and to promote effective grinding as well, to promote more efficient grinding media trajectories, maximize mill throughput with the correct spacing and height of lifter bars, and to minimize liner/lifter and ball damage by avoiding impact onto the mill shell.

This paper presents a short review of mill liner design and its effects on mill performance. Firstly, considerations associated with liner design are briefly described. Then, the effects of mill liner design and operating parameters on mill performance are discussed. The review ends with current trends in establishing appropriate liner design at the operating stage.

2. Considerations in Liner Design

This section begins with a short description of the interaction between mill liner and charge that is a key factor in providing appropriate liner design. Then, liner spacing and shapes that contribute to the improvements in mill performance are discussed.

Mill Liner and Charge Interaction. The effects of the different lifters with the same lifter number/spacing on steady-state charge motion have been predicted using the DEM Millsoft [5]. Figure 1 shows snapshots of the charge motion at a steady state in the 4-m mill for the different lifters. The charge motion of rectangular lifters is almost similar to that of high-low (Hi-Lo) lifters, and a number of free-flying balls are observed for both rectangular and Hi-Lo lifters. Compared with both rectangular and Hi-Lo lifters, trapezoidal lifters with release angles of 20° allow balls to leave from shoulder position earlier and would minimize balls from hitting the mill shell, thereby reducing lifter wear. It is also seen in the mill that triangular lifters with higher leading face angle lower shoulder position of the ball charge, and therefore, it may decrease impact action.

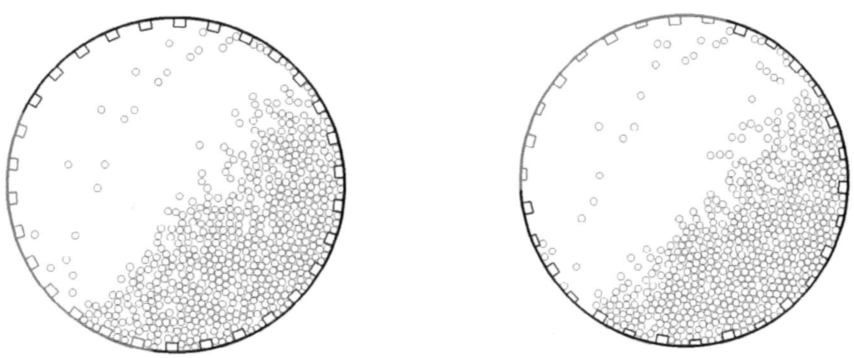

a) Rectangular b) High-Low

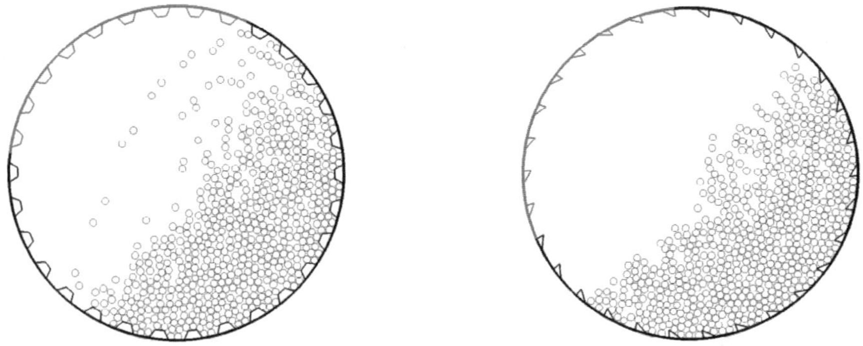

c) Trapezoidal (angle of 20°) d) Triangular

Figure 1. Charge motion of the 4-m mill with different lifter shapes and at the same operating conditions: 70% critical speed, 35% mill load.

Liner Spacing and Shapes. Liner designs depend mainly on various combinations of lifter angles, spacing, height, and mill speed as well. Making changes in lifter spacing and height would modify the charge lifting rate, and thus they affect mill performance. It was found that the different ratios of spacing-to-height of lifter bars in an autogenous mill significantly affect the specific power and mill capacity. Therefore, using the spacing-to-height (S/H) ratio of 4-4.5 could give the optimum specific power and capacity [6].

Furthermore, the Skega A/B (lifter spacing/height) formula, which is an empirical equation for a rubber-lined mill, was formulated and shows that the S/H ratio considerably depends on mill speed [7]. Therefore, the mill could attain a maximum capacity and grinding efficiency at the different S/H ratio. If the lifter spacing is made too widely, it will generate slippage over the shell plate, leading to severe wear. On the other hand, if the lifter distance is spaced too shortly, it would produce material pocket between lifters, resulting in a very high rate of wear on the top of the lifter bars.

Most mill operators have not determined lifter angles at the beginning, and mill suppliers usually decide the angles based on their own experience. The operators consider changing lifter angles to fit their operating conditions just when the liners/lifters underperform or get broken. Powell et al. [1] mentioned some case studies of using good liner designs. For example, the initial SAG mill liner was installed using Hi-Hi with a 6° contact angle. Mill throughput increased by 11% after the face angle had been altered to a 17° angle and then to a 30° angle.

Currently, the large face angle (typically 20° up to 30°) with adequate spacing between lifters is used in the lifter design [8]. The suitable lifter angles could prevent the liner from premature failure and production loss. For mills without packing, computer-program trajectory tools together with DEM have been used for simulating the effects of liner spacing and angles on charge behavior; the outputs showed that mill performance in some cases significantly increases as the results of using wider-spacing and larger lifter-face angles. However, it would decrease as lifter wear becomes larger, and finally the benefits of the initial lifter end as a function of time.

Since the lifter height decreases and the face angle becomes larger because of wear, increasing mill speed for variable-speed mills may lead the ball impacts to directly go down on the toe of the charge and, therefore, compensate for initially wearing lifter to preserve mill performance. Liner wear life, in general, can be increased by increasing lifter height.

3. Effects of Liner Design and Operating Parameters.

The power draw of the mill, which was measured in the wet batch AG mill with different lifter types, decreases exponentially over the milling time [9]. In other words, the power draw of a mill changes significantly with respect to the mass of rock/grinding media remaining in the mill. Optimal liner shapes depend on mill speed, and a lower lifter with a lower face angle is appropriate for mills with higher speeds. The linings leading to the lowest power draw would produce the finest grind and the coarsest grind for those with the highest power draw.

Jonsen et al. [10] also measured torque distribution in detail over one mill revolution at 77% critical speed and found there are obvious peaks in the torque readings when the lifters make a contact with the charge. This indicates that lifter shapes installed in the mill also have an effect on the power draw profiles.

Moreover, mill testing was carried out to measure load cell force of the 1-meter mill installed with the three lifter configurations at the operating conditions: 31% charge (13% ball & 18% rock), 70% critical speed [5]. Figure 2 shows the average power draw of the mill with the different lifters. It appears that the mill power differs slightly for the different lifter shapes with the same face angles of 20°. The mill with rail lifter would relatively require higher power, and with the high-low (Hi-Lo) lifter it does lower power. Even the high (Hi) lifter has the same lifting angles as the Hi-Lo lifter, the use of alternate high and low lifters would markedly affect the internal dynamic of charge in the mill, resulting in a slight change in the mill power draw.

The breakage rate in the laboratory mill with different lifters at the operating conditions: 25% mill charge and 70% critical speed was also determined using a single size of -19+12.7 mm (-3/4+1/2 inch) aggregate. For simplification, the mass fraction, mi(t)/mi(0), of the single size feed remaining in the top size interval versus grinding time is assumed to follow the first order rate. As shown in Figure 3, the breakage rate of the Hi-Lo lifter is highly faster than the first order rate for 4 minutes. Then, it goes down in the range of 4 to 11 minutes and rises at the end. Figure 4 shows that the corresponding size distributions of the mill product is also different with time and at longer grinding times, the rate of production of finer sizes decreases after 8 minutes.

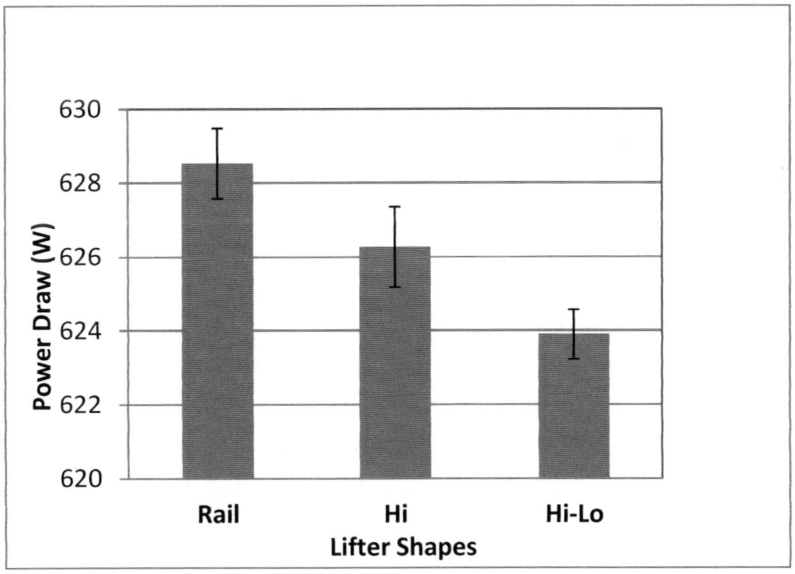

Figure 2. Power draw of the mill with the different lifters, 31% charge: 13% ball & 18% rock, 70% critical speed.

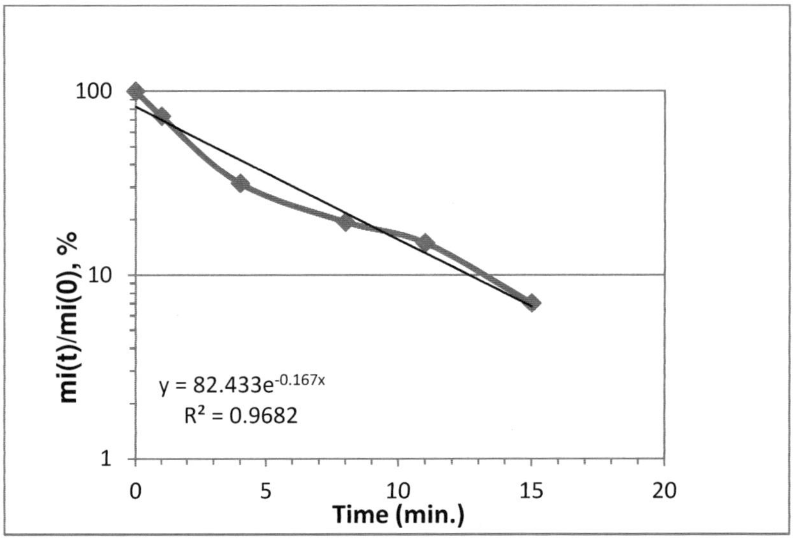

Figure 3. First order plot for the -19+12.7 mm ground in the mill with Hi-Lo lifter at 25% charge, and 70% critical speed.

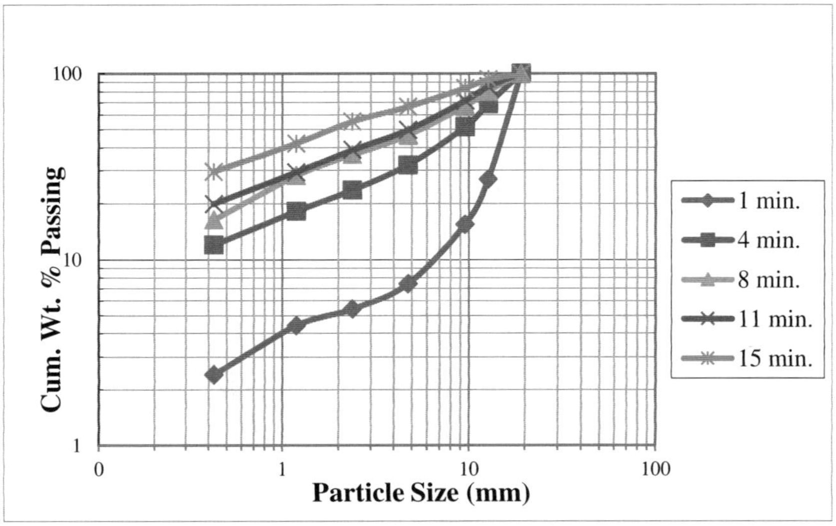

Figure 4. Cumulative size distributions versus time for the -19+12.7 mm ground in the mill with Hi-Lo lifter at 25% charge and 70% critical speed.

4. Establishing Liner Design at Operation

Some mill operators have noticed that mill throughput will systematically change as the liners wear. The wear would lead to significant changes in the shape of lifters, resulting in a significant impact on the grinding efficiency. There is a trade-off between the dual function of liners; designing linier profile to maximize liner life would, on the other hand, decrease the mill power draw, and as a result, the overall grinding rate also decreases. Therefore, it is a good practice for liner development to balance liner life and milling performance.

IOP Conf. Series: Materials Science and Engineering **931** (2020) 012016 doi:10.1088/1757-899X/931/1/012016

The relative wear rates of the different liner components resulting from the measurement of the changes in liner profile over milling time could be used for making a decision to change the liner design [1,11]. For instance, the base section of the liner was found to have a much lower wear rate than the lifter bar. This would be meaningful for the next liners, which should be designed with a little thinner base section. Furthermore, if the wear pattern of the mill liner has been determined, the original design may be made better by redistributing metal from the low wear rate areas to the high wear rate areas. Liner design in the future will be custom-built and not only driven by mill manufacturers but also mill operators. The appropriate design is developed and optimized based on practical experience, operating conditions, and ore types [12,13]. Because of advances achieved in laser-based thickness gauging technology, the tool has been used for measuring the liner wear pattern and rate during service. Combining the data from these scans with the power of DEM simulations and realistic wear testing techniques could provide a more effective way for liner optimization. Therefore, a more efficient liner profile during the liner lifecycle can be determined accordingly.

5. Summary

To establish favorable design, milling management should continue to improve liner design based on useful information from operating experience, ore types, previous good design practice, and supported by sophisticated ball trajectory and DEM tools as well. Due to higher cost and various drawbacks observed from plant trials in industrial mills, the small-scale trials would rather provide important basic information to the design and operation engineers in industrial mills.

The current lifter design utilizes a large face angle with sufficient spacing between lifters. The spacing-to-height (S/H) ratio of 4 at 75% critical speed would work best, and the optimal ratio would change at different mill speeds. The lifters with the same face angles but different configurations would generate different power draw and breakage rates.

Liner design also integrates the wear rates of liners derived from the actual measurement of the changes in liner profile over milling time. Laser-based thickness gauging technology, such as mill mapper, can provide the in-service measurement of the liner wear pattern and rate for determining a more efficient liner profile throughout the liner lifecycle.

References

[1] M. Powell , I. Smit, P. Radziszewski, P. Cleary, B. Rattray, K. Eriksson, and L. Schaeffer, *Selection and Design of Mill Liners, in: S.K. Kawatra (Ed.), Advances in Comminution*, SME, Englewood, Colorado, 2006, pp. 331–376.

[2] R. K. Rajamani, S. Rashidi, and N. Dhawan, *Advances in Descrete Element Method Application to grinding Mills, in: C.* Anderson, R. C. Dunne, and J. L. Uhrie (Eds.), Mineral Processing and Extractive Metallurgy-100 Years of Innovation, SME, Englewood, Colorado, 2014, pp. 117–128.

[3] H. Usman, P. Taylor, and D. E. Spiller, *The effects of lifter configurations and mill speeds on the mill power draw and performance*, in: *AIP Conference Proceedings* 1805 (050001) (2017).

[4] X. Bian, G. Wang, H. Wang, S. Wang, and W. Lv, *Effect of lifters and mill speed on particle behaviour, torque, and power consumption of a tumbling ball mill: Experimental study and DEM simulation, Miner. Eng.* 105 (2017) 22–35.

[5] H. Usman, *Measuring the Efficiency of the Tumbling Mill as a Function of Lifter Configurations and Operating Parameters*, Ph.D. Thesis, Colorado School of Mines, 2015.

[6] R. C. Meaders and A. R. MacPherson, *Technical Design of Autogenous Mills*, *Min. Eng.* 16 (9) (1964) 81–83.

[7] T. K. Moller and R. Brough, *Optimizing the performance of a rubber-lined mill, Min. Eng.* (1989) 849–853.

[8] D. Royston, *Semi-autogenous grinding (SAG) mill liner design and development, Miner. Metall. Process.* 24 (3) (2007) 121–132.

[9] M. S. Powell and L. A. Vermeulen, *The influence of liner design on the rate of production of fines in a rotary mill, Miner. Eng.* 7 (2–3) (1994) 169–183.

[10] P. Jonsén, J. F. Stener, B. I. Pålsson, and H.-A. Häggblad, *Validation of tumbling mill charge-induced torque as predicted by simulations, Miner. Metall. Process.* 30 (4) (2013) 220–225.

[11] M. Yahyaei, S. Banisi, and M. Hadizadeh, *Modification of SAG mill liner shape based on 3-D liner wear profile measurements, Int. J. Miner. Process.* 91 (2009) 111–115.

[12] P. Toor, J. Franke, M. Powell, M. Bird, and T. Waters, *Designing liners for performance not life, Miner. Eng.* 43–44 (2013) 22–28.

[13] M. Yahyaei, M. S. Powell, P. Toor, A. Tuxford, and A. Limpus, *Relining efficiency and liner design for improved plant performance, Miner. Eng.* 83 (2015) 64–77.

IOP Conf. Series: Materials Science and Engineering **931** (2020) 012017 doi:10.1088/1757-899X/931/1/012017

Heat Transfer Analysis on the Tube Type Heat Exchanger with Fin Pitch Variations

A Syuhada[1], D Afandi[1], S E Sofyan[1]
[1]Mechaninal and Industrial Engineering Department, Universitas Syiah Kuala, Darussalam, Banda Aceh, Indonesia

E-mail: syuhada_mech@yahoo.com

Abstract. A heat exchanger is a device usually used to dissipate the heat generated during the mechanical processes. To improve the performance of the heat exchanger, a lot of research has been conducted including the addition of fins on the outer surface. This experimental study examines the heat transfer process on the outer surface of the heat exchanger. The heat exchanger was made of galvanized pipes with an inner and outer diameter of 20 mm and 22 mm. The heat exchanger has a pass length of 30 cm, the spacing between the tube of 6 cm and a total length of 6 m. The spiral fin made of an aluminum plate is mounted to the outer surface of the tube pipe of the heat exchanger. It has a thickness of 0.3 mm. The fin pitch of the heat exchangers was varied to 3 cm, 5 cm, and 7 cm. The heat exchanger was fed with hot water at a constant temperature of 80 ^{o}C and a flow rate of 0.4 L/s. While, the ambient air was blown at the outer surface of the heat exchanger at three different speeds namely 2.4 m/s, 2.8 m/s, and 3.4 m/s. It is found that the highest convective heat transfer rate and convective heat transfer coefficient are yielded by the heat exchanger with a fin pitch of 3 cm namely, 6607.2 W and 235.8 W/m^{2}K, respectively.

1. Introduction

Heat exchangers are a device used to transfer thermal energy between two or more fluids at different temperatures. Heat exchangers are not only used in power generation, petroleum, transportation, cooling (air conditioner, cryogenic, and heat recovery), and manufacturing industries processes, but they also function as key components of many industrial products that available in the market.

Heat exchangers can be classified based on the heat transfer process, amount of fluid involved, and heat transfer mechanism. Heat exchangers can also be classified according to the area of the heat transfer contact. One type of heat exchanger that is commonly used is a compact heat exchanger. Compact heat exchangers have a very high heat transfer surface area to volume ratio and related to their heat transfer coefficient.

The convection heat transfer coefficient (h) greatly influences the heat exchanger performance. The value of the heat transfer coefficient and heat transfer rate in heat exchangers is influenced by the length of the tube, the width of the heat transfer area, and also the type of flow. In the turbulent flow type, the more turbulent flow is, the greater the heat transfer coefficient is attained [1].

The turbulent regime of the working fluid flowing inside the tube might be triggered by creating sharp turns on the heat exchanger. Besides, increasing the flow rate of the working fluid inside the tube induces the Reynolds number to increase and leads the turbulent regime to expand. Introducing fins on the inner surface of the heat exchanger's tube can also generate the turbulent regime of the working fluid. The turbulence regime of the working fluid on the outer surface of the tube can be generated by increasing the flow rate of ambient air crossing through the tube. The addition of fins to the outer surface of the tube also causes turbulence of the airflow to increase [2-5].

Studies on convective heat transfer on the flow inside the pipe with a sharp turn tube have been carried out by Ferianto et al., [6]. Carija et al., [7] analysed the heat transfer on the airside of the flat finned tube and louvered fin heat exchangers. The characteristics of heat transfer of a turbulent flow of a finned heat exchanger with staggered arrangements have been investigated by Bhuyiyan et al., [8]. The heat transfer characteristics of finned tube heat exchangers with several types of fins (plain, louvered, slit, wavy, annular, longitudinal, and serrated fins) were presented by several researchers [9-10]. Gholami et al., [11] examined the heat transfer rate in corrugated tubes and fins.

This study aims to examine the heat transfer rate and the coefficient of the convective heat transfer on the outer surface of a tube-type compact heat exchanger with a sharp turn. The experiment was conducted considering different fin arrangements namely a finless heat exchanger, a heat exchanger with a varied fin pitch of 3 cm, 5 cm and 7 cm.

2. Methodology

2.1. Experimental set up

The heat exchanger (HE) used in this experiment was made of SCH-40 galvanized pipe which has inner and outer diameters are 22 mm 20 mm, respectively. The heat exchanger as a pass length of 30 cm, a tube's spacing of 6 cm and a total length of 6 m. A spiral fin made of aluminum was attached to the outer surface of the tube. The fin has a thickness of 0.3 mm and a height of 10 mm. In this study, four configurations of the heat exchanger were examined including finless, fin pitch 3 cm, fin pitch 5 cm, and fin pitch 7 cm (See Figure 1). Other equipment used includes thermocouple, digital thermometer, fan, hot water drum, hot water inlet pipe, hot water outlet pipe, heater and pump, and stopwatch.

(a) (b) (c) (d)

Figure 1. Heat exchanger configuration (a) finless, (b) heat exchanger with fin pitch 3 cm, (c) heat exchanger with fin pitch 5 cm, (d) heat exchanger with fin pitch 7 cm.

This research was carried out by flowing the hot water into the tube of the heat exchanger. An integrated water heater-pump was used to heat up and to circulate the water into the heat exchanger at a constant temperature of 80 °C and a flow rate of 0.4 L/s. Thermocouples are placed at the inlet and outlet of the heat exchanger to measure the water temperature. A fan placed in front of the heat exchanger was used to circulate the ambient air to the outer surface of the heat exchanger. The air velocity is varied as 2.4 m/s, 2.8 m/s, and 3.4 m/s for each configuration of the heat exchanger (See Fig. 1). Air temperature is measured by placing the thermocouple on the front of the fan and the back of the heat exchanger(10 cm distance from the heat exchanger). Each experiment was conducted for 40 minutes and data was recorded every 5 minutes. The schematic of the test rig is illustrated in Figure 2.

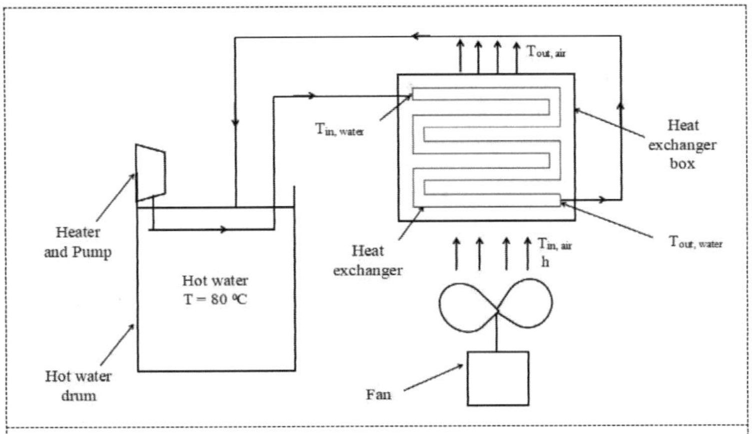

Figure 2. Test rig.

2.2. Heat transfer rate
The rate of heat dissipated by the heat exchanger is calculated as given by Equation 1:

$$Q = \dot{m} \cdot C_p \cdot \Delta T \quad (1)$$

Where Q is the heat transfer rate (W), $\dot{m}$ is the mass flow rate of circulating water (kg/s), C_p is the water specific heat (J/kgK), and ΔT is the water temperature difference between inlet and outlet (K).

2.3. Convective heat transfer coefficient
The coefficient of the convective heat transfer on the airside of the heat exchanger is given as follows:

$$h = \frac{Q}{A \cdot \Delta T} \quad (2)$$

Where h is the convective heat transfer coefficient (W/m^{2}K), Q is the heat transfer rate (W), A is the heat transfer area (m^2), and ΔT is the temperature difference (K), presented by equation 3.

$$\Delta T = T_s - T_\infty \quad (3)$$

Where T_s is the surface temperature of the heat exchanger (K) and T_∞ is the ambient air temperature (K).

In this study, the gradient temperature that occurs in the heat exchanger's pipe wall is ignored. The surface temperature of the heat exchanger is assumed the same as the average temperature of the circulating water as given by equation 4.

$$T_s = \frac{T_{in,water} + T_{out,water}}{2} \quad (4)$$

Where $T_{in\,water}$ is the inlet water temperature (K) and $T_{out,\,water}$ is the outlet water temperature (K).

3. Results and discussion

3.1. Temperature distribution
Figure 3 shows the water temperature distribution for each configuration of the heat exchanger at an air flow of 3.4 m/s. From the experiment, it is found that the outlet water temperature of the finless heat exchanger reduces at an average value of 3.1%. The results reveal that the outlet water temperatures are lower when the spiral fin is introduced to the heat exchanger. This is due to the heat transfer area of the heat exchanger increases. It is found that the outlet water temperatures of the heat exchanger reduced by 4.6 %, 4.8%, and 5% for those with a fin pitch of 7 cm, 5 cm, and 3 cm, respectively.

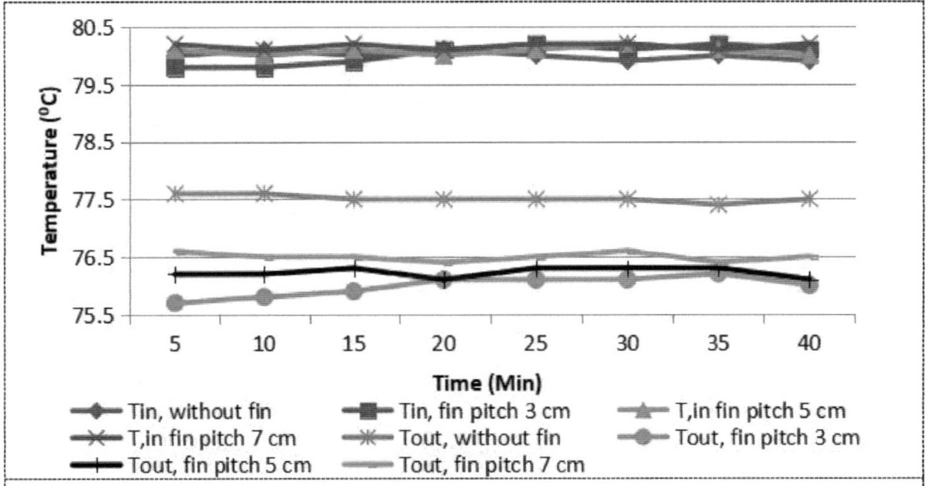

Figure 3. The effect of fin pitch on the outlet water temperature of the heat exchanger at an air velocity of 3.4 m/s.

From Figure 3, it is found that the heat exchanger with a fin pitch of 3 cm yields the best performance amongst others. This section presents the performance of the heat exchanger with a fin pitch of 3 cm, imposing three different ambient air velocity (See Fig. 4). It is found, at At an air velocity of 2.4 m/s, the average water temperature at the heat exchanger outlet decreases by 4.5%. When the air velocity is increased to 2.8 and 3.4 m/s, it is obtained that the average outlet water temperature decrease by 4.8% and 5 %, respectively.

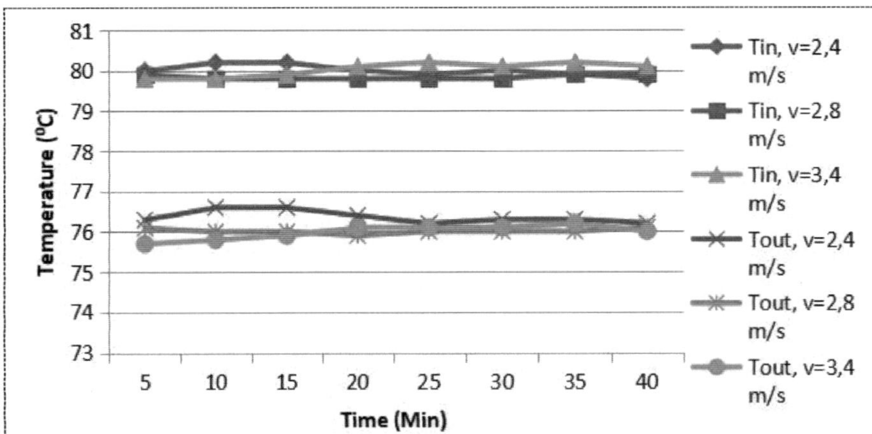

Figure 4. The effect of air velocity on the outlet water temperature of the heat exchanger with a fin pitch of 3 cm.

3.2. Heat transfer rate

The heat transfer rate of the heat exchanger under various configurations is illustrated in Fig. 5. The results reveal that the lowest heat transfer rate is yielded by the finless heat exchanger. The heat transfer rate significantly increases when the fin is mounted on the outer surface of the heat exchanger's tube. This tendency is understandable as increasing the heat transfer area contributes to increasing the heat transfer rate. It is found that the highest heat transfer rate is generated by the heat exchanger with a fin pitch of 3 cm, namely at an average value of 5952,6 W. While, for those with the fin pitches of 5 cm and 7 cm are 5.5 % and 10.6 % lower, respectively. When the air velocity is increased to 2.8 m/s and 3.4 m/s (see Fig. 5a and 5b), it is observed that the same tendency occurs (as that with an air velocity of 2.4 m/s), except for the heat transfer rate which increased with increasing the ambient air velocity on the outer surface of the heat exchanger. This phenomenon occurs due to the air velocity is directly proportional to the coefficient of the convective heat transfer of the ambient air in contact with the outer surface of the heat exchanger. The results reveal that increasing the ambient air velocity on the outer surface of the heat exchanger, from 2.8 m/s to 3.4 m/s has increased the heat transfer rate of the heat exchanger for 7.5 %, 5.5 %, 5.4%, and 6.1%.

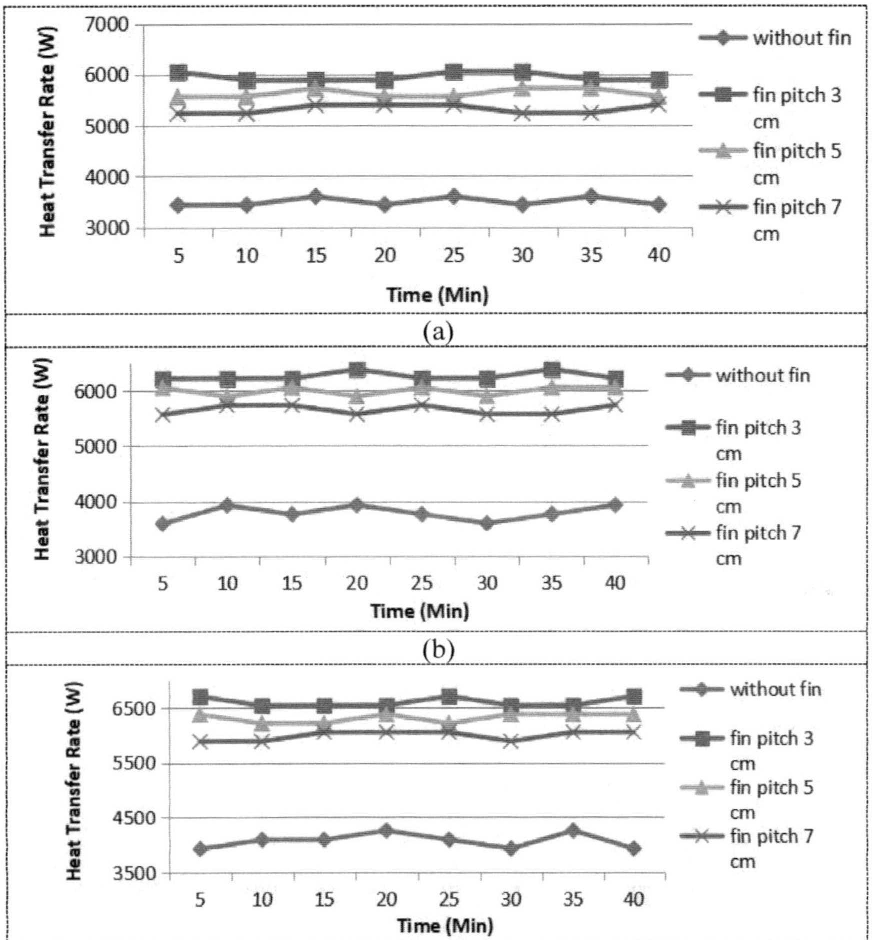

Figure 5. The effect of fin pitch on heat transfer rate of the heat exchanger at an air velocity of (a) 2.4 m/s, (b) 2.8 m/s, (c) 3.4 m/s.

Figure 6 shows the heat transfer rate for the heat exchanger with a fin pitch of 3 cm at three different air velocity namely 2.4 m/s, 2.8 m/s, and 3.4 m/s. It is observed that the heat transfer rate of the heat exchanger is directly proportional to the air velocity. This is because the air velocity affects the Reynolds number. The higher the air velocity, the higher the Reynold number is. This higher Reynold number tends to affect the flow pattern of the ambient air pass through the heat exchanger (i.e turbulence flow). As a result, it is influenced by the heat transfer rate of the heat exchanger. It is found that the average heat transfer rate of the heat exchanger is 5952.6 W at an air velocity of 2.4 m/s. The heat transfer rate increases at an average of 5.2% and 5.6% when the air velocity rises to 2.8 m/s and 3.4 m/s.

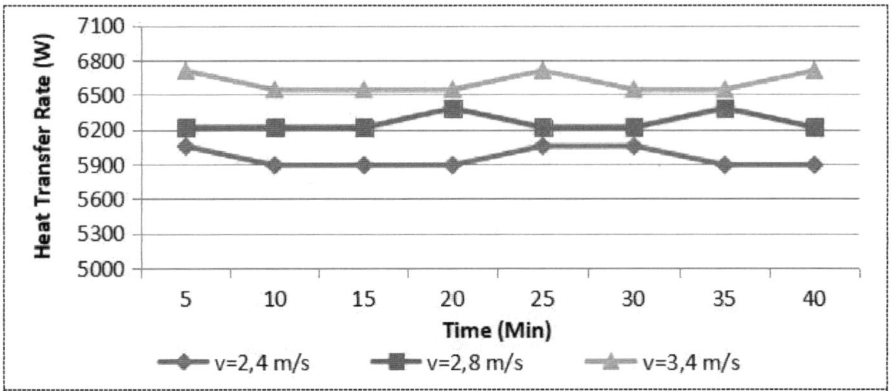

Figure 6. The effect of air velocity on the heat transfer rate of the heat exchangers with a fin pitch of 3 cm.

3.3. Convective heat transfer coefficient

Figure 7 shows the average convective heat transfer coefficients of ambient air in contact with the outer surface of the heat exchanger for studied configurations. The convective heat transfer coefficient is influenced by air velocity, surface area, and temperature difference. The higher the air velocity, the higher the convection coefficient is attained. This is because the air velocity affects the Reynolds numbers. The higher the air velocity, the higher the Reynolds number is generated. At a relatively higher Reynolds number, the fluid flow crossing through the heat exchanger (ambient air) tends to become turbulence. The turbulent flow is one of the factors that influence the convection coefficient. It is found that the highest convective heat transfer coefficient occurs on the heat exchanger with a fin pitch of 3 cm namely 211,75 $W/m^2 \cdot K$, 223,81 $W/m^2 \cdot K$, and 235,85 $W/m^2 \cdot K$, at an ambient air velocity of 2.4 m/s, 2.8 m/s and 3.4 m/s.

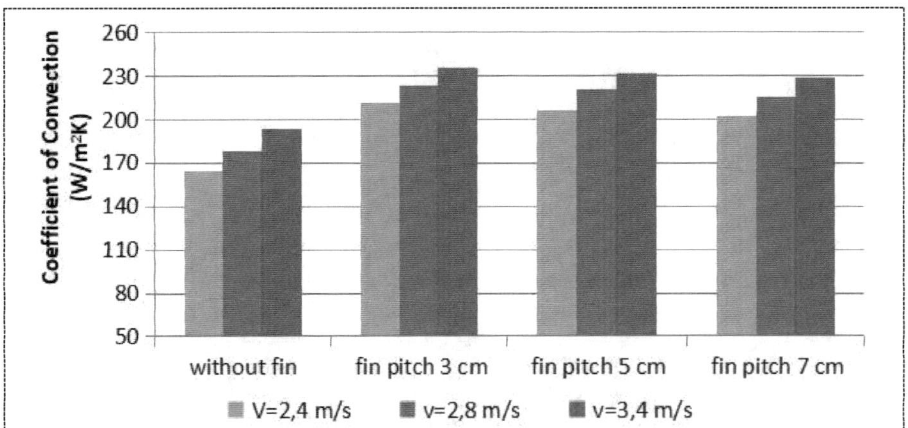

Figure 7. Convective heat transfer coefficient.

4. Summary

The performance of the heat exchanger could be improved by adding fins on its outer surface. The results reveal that the addition of fins to the outer surface of the heat exchanger can significantly increase the heat transfer rate and the convection heat transfer coefficient. The highest heat transfer

IOP Conf. Series: Materials Science and Engineering **931** (2020) 012017 doi:10.1088/1757-899X/931/1/012017

rate and the convective heat transfer coefficient are yielded by the heat exchanger with a fin pitch of 3 cm. These tendencies are affected by the area of the heat transfer and the turbulent flow regime of the air cross through the heat exchanger.

5. References

[1] Shah R K and Sekulic D P 2003 *Fundamental of Heat Exchanger Design* New Jersey John Wiley & Sons, Inc

[2] Hirota M, Fujita H, Nakayama H and Syuhada A 2000 *Heat (Mass) Transfer Characteristic in Serpentine Flow Passage With Sharp Turn* Nagoya Nagoya University

[3] Saputra R D and Putra A B K 2014 Experimental Study Plate Fin Heat Exchanger for Waste Energy Application *Jurnal Teknik POMITS 3* **1-5**

[4] Islam M A and Mozumder A K 2009 Forced Convection Heat Transfer Performance of An Internally Finned Tube *Journal of Mechanical Engineering 40* **54-62**

[5] Kim Y, Kim Y, Kim J and Sin D 2004 Effect of Fin and Tube Alignment on the Heat Transfer Performance of Finned-Tube Heat Exchangers with Large Fin Pitch *Proceeding of International Refrigeration and Air Conditioning* **1-8**

[6] Ferianto 2017 The Effect of Mass Flow Rate On The Effectiveness of A Shell And Tube Heat Exchanger With Double Segmental Baffle *Jurnal Teknik Mesin 5* **67-73**

[7] Carija Z, Francovic B, Percic M and Cavrak M 2014 Heat Transfer Analysis Of Fin-and-Tube Heat Exchanger With Flat And Louvered Fin Geometries *International Journal Of Refrigeration. 45* **160-167**

[8] Bhuiyan A A, Amin M R, Karim R and Islam A K M S 2014 Plate Fin and Tube Heat Exchanger Modeling: Effect Of Performance Parameter For Turbulent Flow Regime *IJAME 9* **1768-1781**

[9] Bhuiyan A A, Amin M R, Karim R and Islam A K M S 2013 Three Dimensional Performance Analysis of Plain Fin Tube Heat Exchangers in Transitional Regime *Applied Thermal Engineering 50* **445-454**

[10] Glazar A, Trp A and Lenic K 2012 Numerical Study of Heat Transfer and Analysis of Optimal Fin Pitch in a Wavy Fin-and-Tube Heat Exchanger *Heat Transfer Engineering 33* **88-96**

[11] Gholami A A, Wahid M A and Mohammed H A 2014 Heat Transfer Enhancement and Pressure Drop for Fin-and-Tube Compact Heat Exchangers with Wavy Rectangular Winglet-Type Vortex Generators *International Communications in Heat and Mass Transfer 54* **132-140**

Thrust Force Analysis of Slotless Permanent Magnet Linear Generator for Wave Energy Converter

Muhammad Tadjuddin[1], Iskandar Hasanuddin[1], Nazaruddin[2] and Amirul Mukminin[3]

[1]Dept. of Mechanical and Industrial Engineering Faculty of Engineering, Universitas Syiah Kuala, Banda Aceh, Indonesia

[2]Dept. of Mathematics Faculty of Science, Universitas Syiah Kuala, Banda Aceh, Indonesia

[3]Dept. Of Electrical Engineering Faculty of Engineering, Universitas Syiah Kuala, Banda Aceh, Indonesia

E-mail: m.tadjuddin@unsyiah.ac.id

Abstract. The WEC (Wave Energy Converter) that uses a permanent magnet linear generator continues to develop. The most important thing to know in designing WEC with a linear electric generator is the thrust required by the linear generator itself. Therefore, this paper discusses and analyses the thrust that occurs in linear electric generators used in WEC systems. Linear electric generators are simulated using FEMM (Finite Element Magnetic Method) software. FEMM simulation is used to analyse the magnitude of the induced magnetic field received by the conductor at a certain position of the stator. The conductor is given a constant current so that it produces thrust force between the stator and translator; the magnitude of thrust force depends on the magnitude of the magnetic field at the position. In the design and analysis, conductors ware given 3A. The maximum thrust force in the linear electric generator design is 79N, and the average force is 75N, force ripple is 1,2%. The design of the permanent magnet linear generator was done, and the WEC system should produce thrust force at least equal to the force that the generator requires.

1. Introduction

The challenge to increase the availability of electrical energy has led to the use of renewable energy that is available in nature. The effort to reduce global warming caused by greenhouse gas emissions from fossil fuel burned in most power engine. The sea wave generator is one of the renewable energy sources found in nature. Ocean wave with potential wave height can be used to generate power, the repulsion force and falling waves are potential energy stayed in the waves so that the thrust force on these waves can be utilized.

In the last few decades, many developed countries have overcome the increasing amount of energy needs using wind or solar energy. Also, renewable energy such as geothermal, biomass, or marine energy is a solution for specific locations. Unlike solar or wind energy, which are present almost everywhere in the world, these forms of energy are somewhat rare, so they can be used in several regions (northern countries use geothermal energy for the production of electrical energy or home heating, agricultural areas use bio-mass for electricity production and areas with coastal shores use ocean energy)[1].

Wave energy converter (WEC) combine with electric generator is the system that took the power of ocean wave energy. An electric generator is a machine that converts mechanical energy into electrical energy. Faraday's law shows that a wire or coil of a conductor in magnetic field that changes with time will produce induced voltage or electromotive force (emf). Faraday's Law is still used as a fundamental calculation for generator design in this modern era.

Wave Energy Converters

Thrust force in ocean waves with their characteristics which have a specific frequency and wave height cannot be directly used to drive a generator. In radial generators, the generator drive needs usually have to go through a rotating shaft. It requires a system that convert potential energy in ocean waves into energy that can be used to rotate the generator shaft. In linear generators, the process of converting potential energy from ocean waves can be used directly because of the characteristics of a linear generator that moves in a straight line with reciprocating movements.

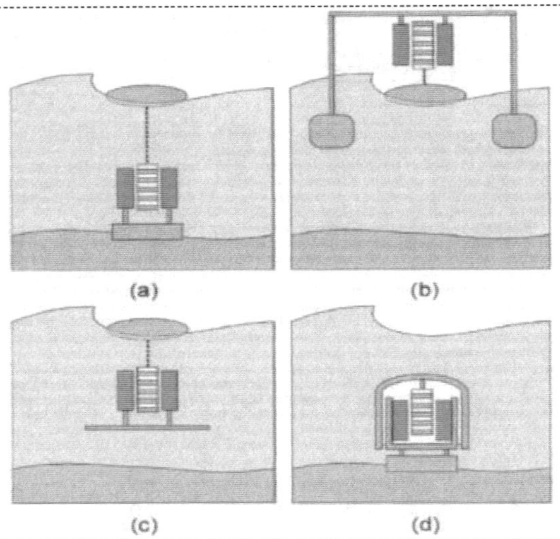

Figure 1. Some variations of WEC for linear generator use [2].

Direct drive system is provided to eliminate mechanical transmission that reduced overall efficiency. The linear electric generator can be used in directly driven by ocean waves energy converter, for example, using float types of the WEC. The simplest using a float connected to the moving generator part (translator), while the stator mounted on the seafloor, as in figure 1 a). Another possible way is to place the stator above sea level using two floaters, positioned in such a way as to give a half-wave distance between the floaters and the translator, as in Fig. 1 b). Another solution is to let the stator float above sea level 1 c). or by placing the entire structure in a closed cylinder, such as Figure 1 d). Depending on the amount of water that is above the device, the air pressure in the cylinder will cause up and down movements of the translator to produce electric energy [2].

Wave Energy Converters (WEC) is used to change the direction of the force on ocean waves so that it matches the required generator drive. WEC can use to drive two types of electric generators, namely rotary generators or linear generators. Unlike a rotary generator, a linear generator can be directly driven by movement at the WEC because of the vertical movement of ocean waves while a rotary generator requires a transmission system to be driven [3]. Most Wave Energy Converters (WEC) devices use mechanical or hydraulic systems to transmit ocean wave power to linear generators, based on the device with a transmission system that uses many inefficient gearboxes, due to energy losses in the gearbox. In

IOP Conf. Series: Materials Science and Engineering **931** (2020) 012020 doi:10.1088/1757-899X/931/1/012020

general, the linear generator in the WEC operated by moving the translator up and down, adjusted to the wave motion at the WEC in contact with ocean waves, for example, buoy [4]

Electric Linear Generator

A linear generator is an electric machine that can produce electricity by converting it from of linear motions. The advantage of linear generators is the use of movements that can directly use the kinetic motion of generators that move linearly without going through mechanical transmission [5].

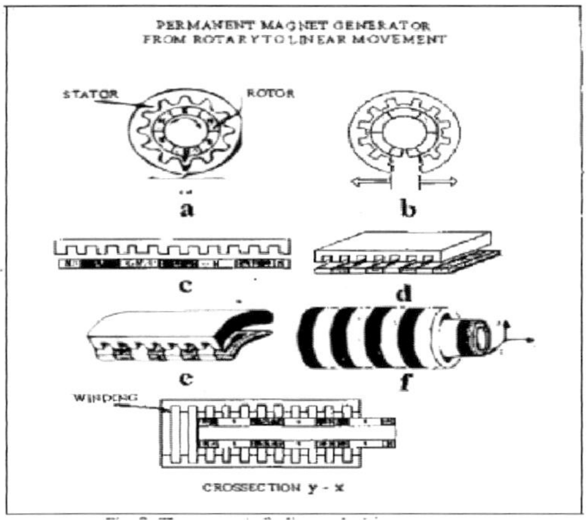

Figure 2 . The concept of a linear generator [5]

Fig. 2 shows how a linear electric generator is formed from a rotary generator. When the rotary generator stator is cut and opened (Fig. 1 a and 1 b), a flat side topology is obtained for the stator for the linear generator. The same process applies to the rotor and as a result, a magnetic configuration will be reached (fig. 1 c and 1 d). The stator structure is flat, and the magnet is then converted to cylindrical shape (figures 1 e and 1 f). The coil is also rolled and inserted into the stator slot. And got a concept for a linear generator[6].

The working principle of a linear generator is not much different from other conventional generators, in general, the process of the magnetic induction of the coil is done to produce voltage, but to generate magnetic flux can be done by designing the windings or using permanent magnets on the rotor. In Linear Generators no current or excitation voltage is needed on the rotor because permanent magnets are used to generate magnetic flux, it can be explained simply, the voltage will be generated in the conductor (stator winding) if the conductor moves in a magnetic field. This certainly will increase efficiency because the flow of conversion of mechanical energy into electricity can be done directly without external generator excitation.

Design Topology Modular Linear Generator

In this paper, a linear generator is designed in an independent modular model, as shown in figure 3. It can be used in one separate generator module or can also be assembled in the form of a multi-module. Multi-module type, as shown in figure 4 assembly in series or parallel.

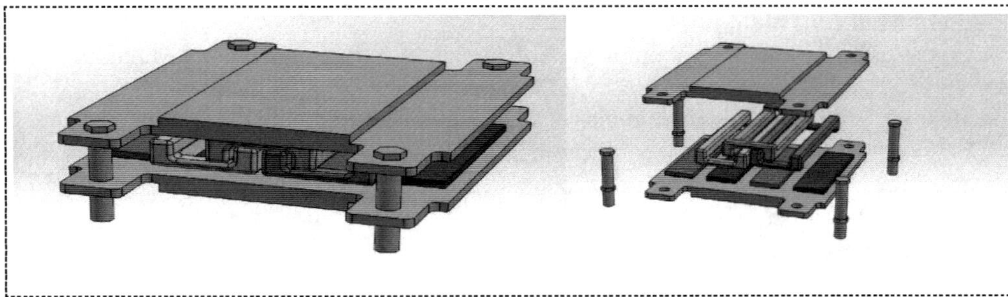

Figure 3. Linear design of a modular generator in one module

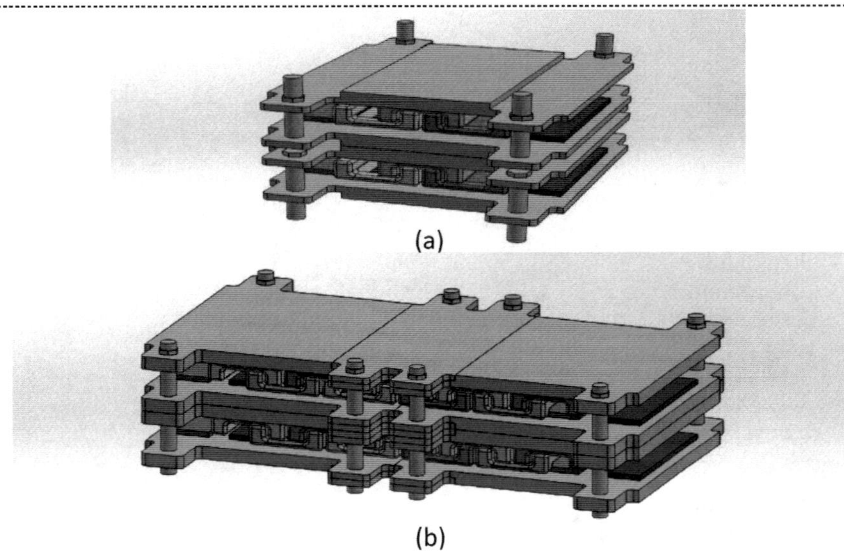

Figure 4. Modular design a. multilevel connection b. elongated connection

The linear electric generator design in this paper is designed with a coreless type, there is absence of laminating iron or iron core in the stator winding. The conductor winding is fixed by resin. Because of the absence of laminating iron or iron core in this design, this generator types are also called a slotless generator. Coreless type is better than the slotted type linear generator in terms of force ripple, because there is no interaction force between magnet and slotted iron. The linear electric generator parameters for one module are in the following table:

Table 1. Linear generator parameters for one module

Parameters	**Amount**	**Unit**
Magnet dimention	150x40x8	mm
Type of Magnet	-	NdFeb 40
Magnet per module	4	module
Space between magnets	20	mm
Number of winding	100	Turn
Diameter wire	1	mm
Air gap	1	mm

Finite Element Method Magnetics Simulation

In this paper, the simulation is carried out using Software Finite Element Method Magnetics version 4.2 (FEMM 4.2). The topology and linear design of the generator are depicted in a 2-dimensional (2D) form, and parameters given to the component material used are as shown in table 1 above.

In figure 5 shows the linear generator topology in 2 dimensions, there are two sides of back iron with a thickness of 15 mm, on each side, there are four magnets, the type of magnet used is NdFeb 40, with air-gap of 1 mm. The thrust force on the translator of a linear generator is obtained by injecting the current in the windings.

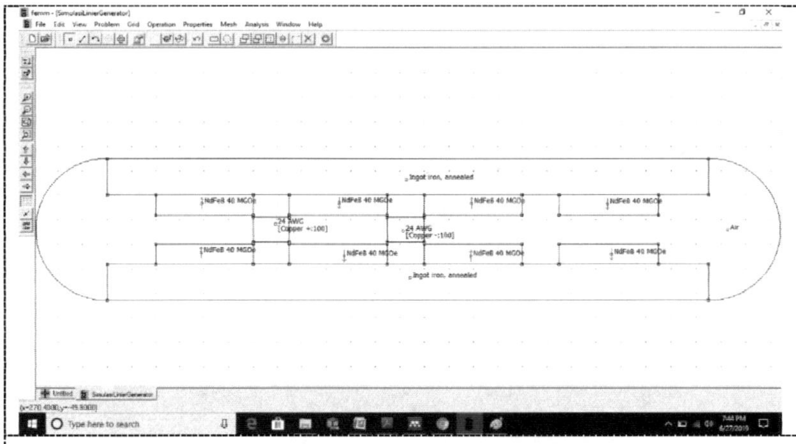

Figure 5. 2D topology design in FEMM 4.2

2. Result and Discussion

The flux density is obtained by running a simulation. The magnetic field lines can be measured for strength, as in Figure 6 shows the distribution of the magnetic flux density in a linear generator module.

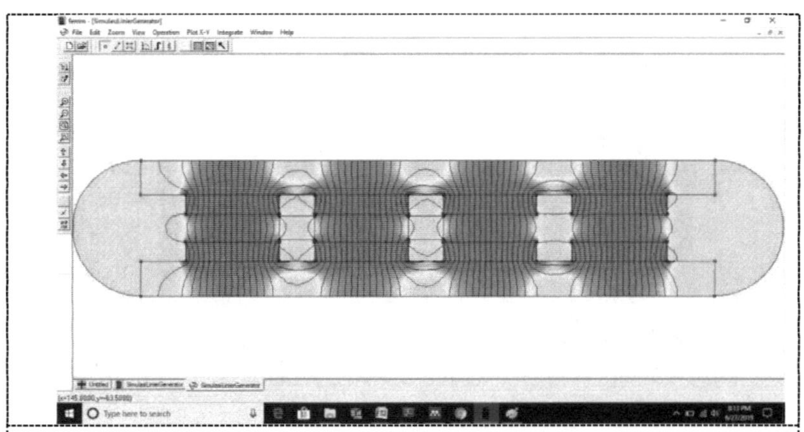

Figure 6. Distribution of Flux in simulations

The flux density is shown in Figure 7 below, the flux density of each part of the linear generator in simulation has a certain color, where the color identifies the flux density.

Flux density results are obtained as shown in Figure 8, the maximum flux is 0.697 T. Flux density can be interpreted as the strength of the magnetic field used. In this paper, the type of magnet used is grade 40 Neodymium Feron Boron (NdFeB 40). The results shown in Figure 8 are the results of testing in one of the generator phases.

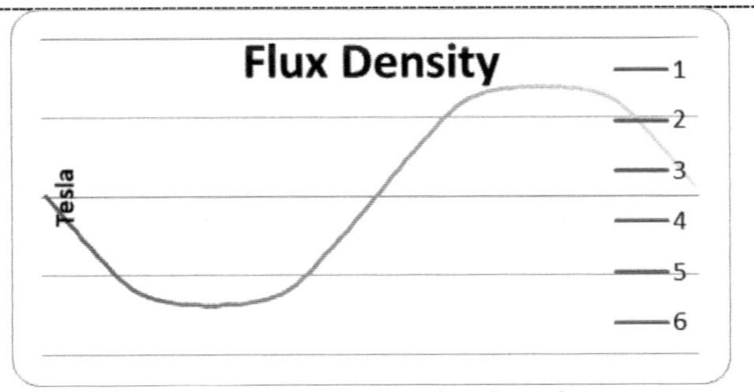

Figure 7. Phase A Flux Density

The purpose of this study is to measure the power needed to drive a linear generator. The required thrust force can be obtained by injecting a current into the windings. In this study the current injected during the simulation was 3.5 A, and Thrust Force was obtained in each phase, as shown in Figure 9 below.

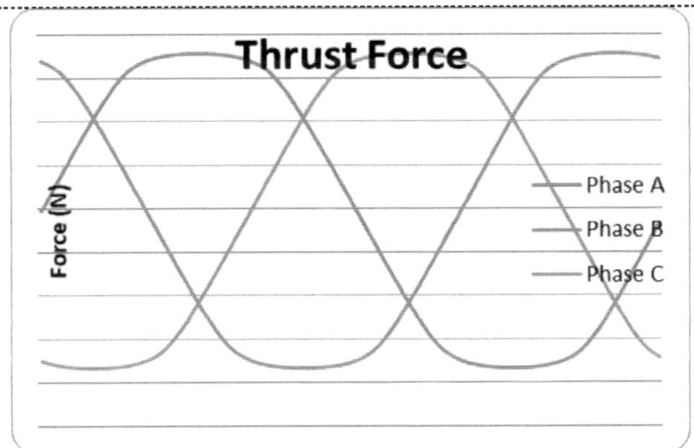

Figure 8. Thrust Force in each phase

Fig. 9 shows when each phase is injected with a current of 3.5 A, a maximum force of 35.7 N. and a maximum average thrust force of 79 N are obtained as shown in Figure 10 below.

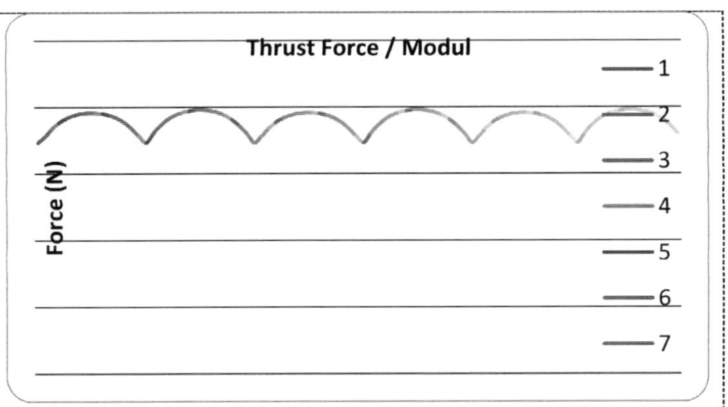

Figure 9. Thrust Force on one module

IOP Conf. Series: Materials Science and Engineering **931** (2020) 012020 doi:10.1088/1757-899X/931/1/012020

3. Conclusion

Design and simulation of a linear generator with a 3-phase coreless configuration using permanent magnets have been done. The linear generator was in design modularwise for wave energy coveter, where the number of turns per phase coil is 100 turns. Reaction force carried out on this generator is an average of 79 Newtons, at a fix current injected of 3.5 Amperes. The data is retrieved only for a set of 3 phase winding with star connections. The average thrust force is 75N, with 1.2% force ripple. In WEC the output of voltage depend on wave height and period which correspond to linear generator slider or translator speeds in direct driven system.

Acknowledgment

This research is funded by RISTEKDIKTI on research grant of PTUPT 2019, under LPPM and fully support of Laboratorium Design and Manufacture, Syiah Kuala University

References

[1] Cruz Joao, *Green Energy, and Technology – Ocean Wave Energy, Current Status and Future Perspectives*. Springer Netherlands, 2008.
[2] C. A. Oprea, C. S. Martis, F. N. Jurca, D. Fodorean, and L. Szabó, "*Permanent Magnet Linear Generator for Renewable Energy Applications : Tubular vs. Four-Sided Structures,*" pp. 588–592, 2013.
[3] B. Drew, P. A.R, and S. M.N, "*A Review of Wave Energy Converter Technology*," *Proc. Inst. Mech. Eng.*, vol. 223, pp. 887–902, 2009.
[4] N. A. Mohd Zamri, T. Ibrahim, and N. Mohd Nor, "*Direct-Drive Linear Generator Designs with Aluminium Spacer and Alternate Slot Winding for Wave Energy Conversion System*," *Int. J. Adv. Sci. Eng. Inf. Technol.*, vol. 7, no. 4, p. 1282, 2017.
[5] I. Boldea and S.A. Nasar, "*Linear electric actuators and generators*," *IEEE Trans Energy Convers.*, vol. 14, pp. 712–717, 1997.
[6] W.P.Hew, J. Jamaluddin, M. Tadjuddin, and K. M. Nor, "*Fabrication and Testing of a Linear Electric Generator for Use with a Free-Piston Engine*," *IEEE*, 2003.
[7] J. F. Gieras, R.-J. Wang, and M. J. Kamper, "*Axial Flux Permanent Magnet Brushless Machines*". Dordrecht: Springer Netherlands, 2008.

Biomechanical Posture Assessment of Salted Fish Industry Workers in West Aceh

Husni[1], I Hasanuddin[1], K Hadi[2], Akram[1], T Firsa[1]
[1]Mechaninal and Industrial Engineering Department, Universitas Syiah Kuala, Darussalam, Banda Aceh, Indonesia
[2]Industrial Engineering Department, Universitas Teuku Umar, Aceh Barat, Indonesia

E-mail: husniusmn@unsyiah.ac.id

Abstract. Salted fish is dried fish that is preserved by adding salt as the preservative medium. The process of production starts with cleaning, salting, boiling, drying, sorting, and packing. The length and complexity of the process lead to an increase in the risk of MSDs and posture complaints among the workers. This study aimed to find out the factors that caused the complaints with the biomechanical posture method. Six subjects were measured for the tension at erector spinae and the moment at (lumbar 5 sacrum 1) L5-S1 vertebral joint created by body segments and load, the calculation of the workers' body segment weight, the distance of the body's center point to the lumbar 5 sacrum 1 (L5-S1), and biomechanics at L5-S1. The largest body segment weight was found at the cleaning station, which dealt with the process of manually lifting the fish, with a bodyweight value of 70 kg, W_{trunk}: 350 N, W_{head}: 58.8 N, W_{arm}: 35.7 N and W_{load}: 300 N. The farthest distance from the body's center point to L5-S1 was found at the drying station, which deals with the process of drying the fish, with values of the distance of head (d_{head}): 54.78 cm, distance of trunk (d_{trunk}): 59.26 cm, distance of arm (d_{arm}): 67.39 cm and distance of box (d_{box}): 86.11 cm. The largest moment in force on L5-S1 was found at the packing station, which deals with the fish transferring and packaging process, with the value of the moment of force at 5351.86 N and 4215.28 N. Therefore, it is suggested that the salt fish industry finds solutions, re-designing the work station, decreasing the lifting load, and producing proper equipment to assist workers during labor to improve their safety and health.

1. Introduction

Salted fish is one of the commodities in Aceh province, Indonesia, with high production in the district of West Aceh, an area on the coastline. Most of the residents in the area work as fishermen. Big catches cause the fishermen to find solutions so that the excessive fish do not rot. Hence, the alternative is to make salted fish.

Salted fish is dried fish preserved by adding salt as the preservative medium. It is processed from marine fish, which is then preserved traditionally. The traditional method of preservation aims to reduce the water content in the body of the fish, so as not to provide an opportunity for bacteria to grow. Besides, salted fish has been found to contain a rich source of proteins, approximately 80-85% of protein, which would be a good solution for consumers [1].

The work process carried out at the place of business of making salted fish, starting from cleaning, salting, boiling, drying, sorting, and packing of the fish. This can increase the risk of musculoskeletal complaints. The boiling work process is the most exhausting work among the routines of making salted fish. This is due to the static standing posture in the boiling work, the work facilities that are not in accordance with anthropometry, direct contact with the stove and the hot steam, and the strong, pungent smell of the fish.

According to Tarwaka and Sudiajeng [2], subjective complaints of skeletal muscle disorders with static standing posture are commonly reported at 54.2% in the lower neck, 33.3% in the right shoulder, 50% in the back, 87.5% in the waist, 45.8% in the right and left thighs, 58.3% in the right and left knees, and 33.3% in the right and left calves. Meanwhile, subjective complaints of skeletal muscle disorders with dynamic standing posture are commonly reported at 37.5% in the lower neck, 62.5% in the right shoulder, 50% in the back, 87.5% in the waist, 41.7% in the right and left thighs, 16.7% in the right and left knees, and 54.2% in the right and left calves. These skeletal muscle disorders affect the capacity of workers to carry out their work.

In 1994, 705,800 cases (32%) of all cases in the United States which occurred due to overexertion or repetitive motions, including 367,424 cases of back pain, were caused by overexertion in lifting and 65% of the cases affected the back of workers. From 93,325 cases which were due to overexertion in pushing or pulling objects, about 52% of them affected the back. As is usually the case in the industry, the position and procedures of work are often not well designed so that the operational performance is not optimal. Besides, the ineffective working conditions accelerate fatigue and cause many complaints, pain, and injury to the limbs in the short and the long term to the workers [3][4]. A similar case is experienced by workers in the salted fish industry. Improper work positions can cause interference with the body parts of the workers. This is very likely to cause injury if done inappropriately.

2. Methodology

At this stage, it is necessary to know what factors cause a problem, which is necessary for the research so that the right solution can be offered to solve the problem. The problems found in this industry were: mapping of salted fish craftsmen, MSDs and posture factors, and identification of complaints from the workers using the Nordic Body Map questionnaire. Data were collected for a month at UKM (*Usaha Kecil dan Menengah* or Small Medium Industry) Stefen Aluy, Meulaboh, West Aceh, from six subjects (i.e. workers) who worked eight hours a day in a week. They were done through interviews and surveys. The interviews were conducted directly during their work, using a video camera. Calculations were also conducted on the workers' body segment weight, the distance of the body's center point to the lumbar 5 sacrum 1 (L5-S1), which is the distance of head (d_{head}), the distance of the trunk (d_{trunk}), the distance of the arm (d_{arm}) and distance of the box (d_{box}), and the touque and moment of the arm at L5-S1 in the static position. Figure 1 shows the erector spinae with a moment of the arm of 6 cm from the L5-S1 joint center [5]. The moments of force on the activities experienced extreme muscle stress on the L5-S1 and are classified as dangerous based on NIOSH's standard of 3400 N for maximum load lifting [6].

$$Torque = Force(F_m) \times MomentArm \tag{1}$$

$$\begin{gathered}\sum T_{L5-S1} = 0 \\ 0 = (F_m)\times(6cm) - \left[(W_{head})\times(d_{head}) + (W_{trunk})\times(d_{trunk}) + (W_{arms})\times(d_{arms}) + (W_{box})\times(d_{box})\right]\end{gathered} \tag{2}$$

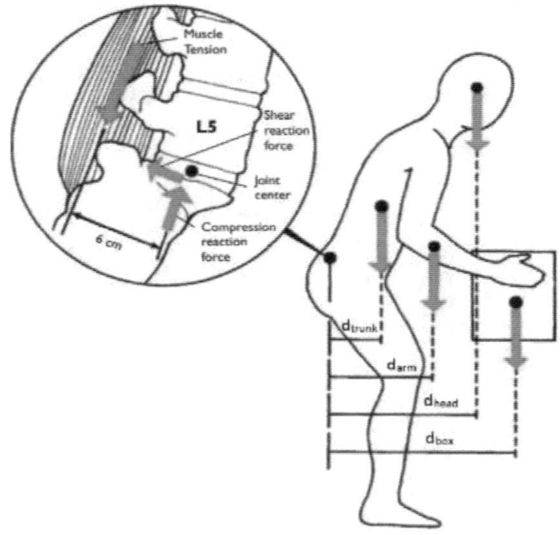

Figure 1. Illustration of workers lifting the box [5].

3. Results and discussion

Based on surveys conducted with workers, using the Nordic Body Map questionnaire [7] the initial data was obtained in the form of a questionnaire distributed to six workers. Table 1 shows the description of workers' complaint percentages obtained through the questionnaire. Approximately 45.24%, or about half, of the informants, felt "painful" during work. They informed that the most pain was felt in the neck, waist, back pain, and others (see Table 2).

Table 1. Recapitulation of workers' complaint percentages.

No.	Parameter of Complaint	Percentage
1	Not painful	21.43
2	Slightly painful	33.33
3	Painful	45.24

From Table 2, for instance, four out of six workers stated "painful" in the neck and waist while working, while two workers stated "very painful". This indicates that the work carried out by the worker was forceful and required improvement of the work system to minimize the risk of injury to the body, especially in the neck and waist.

Table 2. Workers' complain percentage.

No.	Type of complaint	Level of complaint			
		Not painful	Slightly painful	Painful	Very painful
		%	%	%	%
1	Upper neck pain/stiffness	16,7	50,0	33,3	-
2	Lower neck pain/stiffness	-	50,0	50,0	-
3	Left shoulder pain	33,3	-	66,7	-
4	Right shoulder pain	-	33,3	66,7	-
5	Left upper arm pain	16,7	50,0	33,3	-

6	Back pain	-	50,0	50,0	-
7	Right upper arm pain	33,3	-	66,7	-
8	Waist pain	33,3	-	66,7	-
9	Buttock pain	-	33,3	66,7	-
10	Hip pain	16,7	50,0	33,3	-
11	Left elbow pain	16,7	50,0	33,3	-
12	Right elbow pain	-	50,0	50,0	-
13	Left lower arm pain	33,3	-	66,7	-
14	Right lower arm pain	-	33,3	66,7	-
15	Left wrist pain	16,7	50,0	33,3	-
16	Right wrist pain	16,7	50,0	33,3	-
17	Left hand pain	-	50,0	50,0	-
18	Right hand pain	83,3	16,7	-	-
19	Left thigh pain	83,3	16,7	-	-
20	Right thigh pain	83,3	16,7	-	-
21	Left knee pain	16,7	50,0	33,3	-
22	Right knee pain	16,7	50,0	33,3	-
23	Left calf pain	-	50,0	50,0	-
24	Right calf pain	33,3	-	66,7	-
25	Left ankle pain	-	33,3	66,7	-
26	Right ankle pain	16,7	50,0	33,3	-
27	Left foot pain	-	50,0	50,0	-
28	Right foot pain	33,3	-	66,7	-

Meanwhile, Figure 2 shows that the most felt pain by the workers was "painful". This indicates that the workers' working condition was not safe and could result in danger to their health. In some cases, when this situation occurred, the industry could fabricate equipment to assist and ease the pain of workers while working [8].

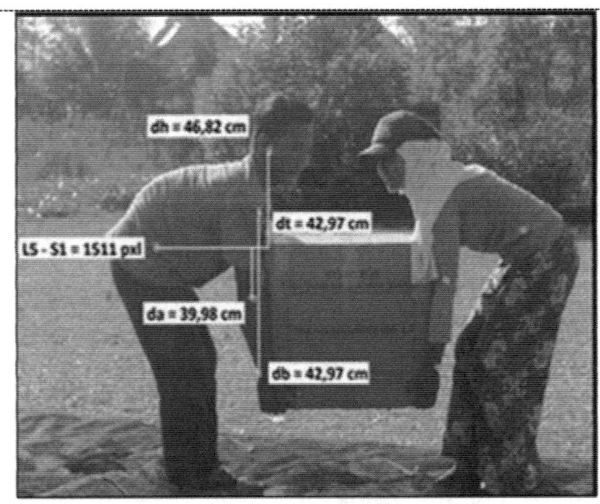

Figure 2. The process of moving fish with a moment force of 5351.80 N.

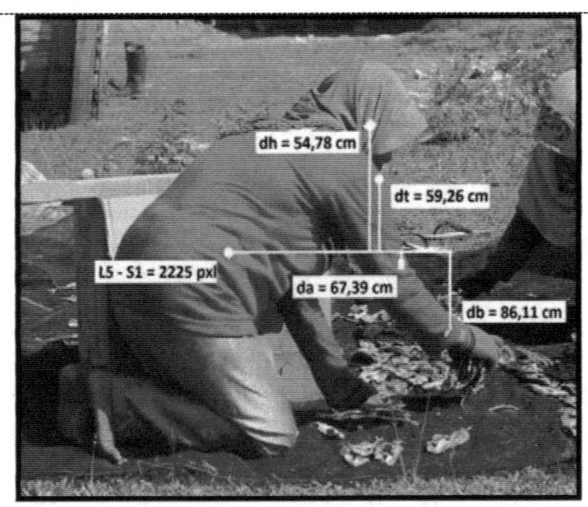

Figure 3. The fish packaging process with a moment force of 4215.28 N

Calculating the distance of the body's center point to the lumbar 5 sacra 1 (L5-S1), and biomechanics at L5-S1. Before the analyses, the data must be normal, uniform, and sufficient. Therefore, it is necessary to conduct a data normality test, data uniformity test, and data adequacy test.

3.1. Body segment weight

Body Segment Weight (BSW) for each worker in this study is shown in Tables 3-7. Meanwhile, the BSW for workers in the salted fish industry was calculated using the equation below [9].

Table 3. Sorting workstation.

No	Subject	Work Posture	Body Weight (Kg)	W Trunk (N)	W Head (N)	W Arm (N)	W Load (N)
1	Subject 1	Sorting	55	269.77	46,2	28,05	5
2	Subject 1	Lifting the Box	55	275	46,2	28,05	250

Table 4. Cleaning workstation.

No	Subject	Work Posture	Body Weight (Kg)	W Trunk (N)	W Head (N)	W Arm (N)	W Load (N)
1	Subject 2	Lifting the box	45	225	37,80	22,95	250
2	Subject 2	Cleaning	45	225	37,80	22,95	5

Table 5. Boiling Workstation

No	Subject	Work Posture	Body Weight (Kg)	W Trunk (N)	W Head (N)	W Arm (N)	W Load (N)
1	Subject 6	Lifting the box fish	66	330	55,44	33,66	125
2	Subject 3	Boiling	70	350	58,8	35,7	5

Table 6. Drying workstation

No	Subject	Work Posture	Body Weight (Kg)	W Trunk (N)	W Head (N)	W Arm (N)	W Load (N)
1	Subject 4	Lifting the Box	45	225	37,8	22,95	125
2	Subject 5	Sorting fish	56	280	47,04	28,56	5

Table 7. Packing workstation

No	Subject	Work Posture	Body Weight (Kg)	W Trunk (N)	W Head (N)	W Arm (N)	W Load (N)
1	Subject 3	Lifting the Box	70	350	58,8	35,7	300
2	Subject 6	Packing	66	330	55,44	33,66	5

3.2. Calculation of distance from body segment weight to L5-S1

The distance from body segment weight to L5-S1 for workers in the salted fish industry was performed using the following calculation using the still picture to measure the distance in pixel and convert the distance to cm.

3.2.1. Subject 1 (Sorting workstation in Figure 4)

IOP Conf. Series: Materials Science and Engineering **931** (2020) 012021 doi:10.1088/1757-899X/931/1/012021

Calculation of dt (distance from L5-S1 to trunk segment weight)

$$dt = (3085 - 2873) \times 0.059$$
$$= 212 \times 0.059$$
$$= 12.51cm$$

Calculation of dh (distance from L5-S1 to head segment weight)

$$dh = (3085 - 1178) \times 0.059$$
$$= 324 \times 0.059$$
$$= 19.12cm$$

Calculation of da (distance from L5-S1 to arm segment weight)

$$da = (3085 - 2851) \times 0.059$$
$$= 234 \times 0.059$$
$$= 13.81cm$$

Calculation of db (distance from L5-S1 to load segment weight)

$$db = (3085 - 2585) \times 0.059$$
$$= 500 \times 0.059$$
$$= 29.5cm$$

Figure 4 shows the results of the calculation of the distance from body segment weight to L5-S1 from Subject 1 at the sorting work station. The details of the calculation of the distance from body segment weight to L5-S1 for workers at each work station are shown in Table 8.

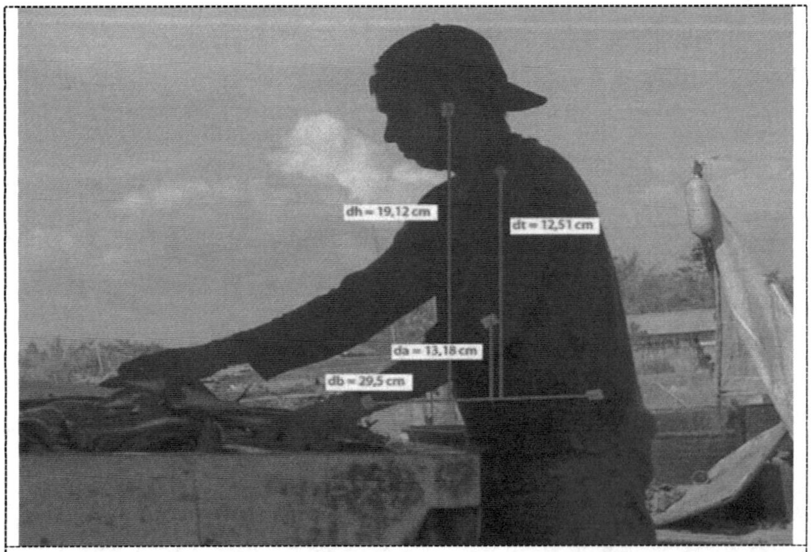

Figure 4. Sorting work station measurement distance from body segment weight to L5-S1.

Table 8. Recapitulation of calculation of the distance from body segment weight to L5-S1.

No.	Work station	Subject	Work Posture	Distance from body segment weight to L5-S1 (cm)			
				D_{head}	D_{trunk}	D_{arms}	D_{box}
1	Sorting	1	1	36,24	31,27	31,16	35,91
		1	2	19,12	12,51	13,81	29,50
2	Cleaning	2	1	27,55	18,96	22,81	33,77
		2	2	37,29	24,82	27,79	54,17
3	Boiling	6	1	7,79	11,51	14,87	27,08
		3	2	37,09	25,48	32,55	48,56
4	Drying	4	1	8,60	14,23	24,02	40,62
		5	2	31,94	35,85	48,26	64,34
5	Packing	3	1	46,82	42,97	39,98	42,97
		6	2	54,78	59,26	67,39	86,11

3.3. Calculation of moment force on L5-S1

The moment of force calculation on L5-S1 was done to find out the muscle stress level that occurred at L5-S1 during salted fish processing work. Calculation using formula No.2 of moment of force for Subject 1 at the sorting work station is as follows.

$$\sum T_{L5,S1} = 0$$

$$(F_m \times E) - \sum (W \times d) = 0$$

$$(F_m \times 6) - \{(275 \times 12.51) + (46.2 \times 19.12) + (28.05 \times 13.81) + (5 \times 29.5)\} = 0$$

$$(F_m \times 6) - 4858.46 = 0$$

$$F_m = \frac{4858.46}{6}$$

$$F_m = 809.74$$

The details of the calculation of moment of force in workers in the salted fish industry at each work station can be seen in Tables 9-13.

Table 9. Moment of force at sorting workstation

No.	Subject	Work Posture	Bodyweight (kg)	Moment of force (N)
1	Subject 1	Sorting	55	809.74
2	Subject 1	Lifting the Box	55	3354.18

Table 10. Moment of force at cleaning workstation

No.	Subject	Work Posture	Bodyweight (kg)	Moment of force (N)
1	Subject 2	Lifting the box	45	2378.90
2	Subject 2	Cleaning	45	1317.12

Table 11. Moment of force at boiling workstation

No.	Subject	Work Posture	Bodyweight (kg)	Moment of force (N)
1	Subject 6	Lifting the box fish	66	813.80
2	Subject 3	Boiling	70	2052.41

Table 12. Moment of force at drying workstation

No.	Subject	Work Posture	Bodyweight (kg)	Moment of force (N)
1	Subject 4	Lifting the Box	45	1525.93
2	Subject 5	Sorting fish	56	2206.74

Table 13. Moment of force at packing workstation

No.	Subject	Work Posture	Bodyweight (kg)	Moment of force (N)
1	Subject 3	Lifting the Box	70	5351.80
2	Subject 6	Packing	66	4215.28

4. Conclusion

Based on the calculation of the moment of force at each work station, excessive moments of force were found at the packing station, namely on the fish transferring and the packaging process, with the value of 5351.86 N and 4215.28 N respectively. The moments of force in the activities experienced extreme muscle stress in L5-S1 and are classified as dangerous based on NIOSH's standard of 3400 N for maximum weight lifting. Therefore, it is suggested that the salt fish industry finds solutions such as work station be re-designed and the lifting load be decreased, and producing equipment to assist workers during labor and hence improve their safety and health.

5. Acknowledgement

In carrying out this research, the authors express their gratitude to the Work System Design and Ergonomics Laboratory, Department of Mechanical and Industrial Engineering, Faculty of Engineering, Universitas Syiah Kuala, for the facilities provided in supporting this research. This research is part of the postgraduate thesis. Authors' appreciations are addressed to the UKM (*Usaha Kecil dan Menengah* or Small Medium Industry) Stefen Aluy, Meulaboh, West Aceh, for consenting to collect data at the site.

6. References

[1] Yulisa N Y, Asni E and Azrin M 2014 Uji formalin pada ikan asin gurami di pasar tradisional Pekanbaru [Formalin test on salted gourami in Pekanbaru traditional market]. *J Online Mhs JOM Bid Kedokt, Vol. 1(2)* **1–12**

[2] Tarwaka S H and Sudiajeng L 2004 *Ergonomi untuk keselamatan, kesehatan kerja dan produktivitas* [Ergonomics for safety, occupational health and productivity] Surakarta UNIBA

[3] Erwan F, Iqbal M and Hasanuddin I 2018 The analysis of nursing's work systems in relation to burnout syndrome (A case study: nurses in RSUDZA, BLUD RSIA, and RSUD Meuraxa, Banda Aceh, Indonesia). *IOP Conf. Series: Materials Science and Engineering, Vol. 453* **1-8**

[4] Hasanuddin I, Fahrizal R and Asmadi D 2019 Analysis of transport workers' postures in the loading process of manual material handling activities by using the photogrammetric method. *IOP Conf. Series: Materials Science and Engineering, Vol. 523* **1-10**

[5] Hall S J 2012 *Basic biomechanics, 6th Ed.* New York McGraw-Hill

[6] Bridger R S 1995 *Introduction to ergonomics* New York McGraw-Hill Inc.

[7] Wilson J R and Corlett E N 1995 *Evaluation of human work: A practical ergonomic methodology* London Taylor & Francis

[8] Salam R, Iqbal M and Hasanuddin I 2018 Desain dan analisis artificial exoskeleton pada prajurit TNI [Design and analysis of artificial exoskeleton in TNI soldiers] *Jurnal Optimasi Sistem Industri (JOSI) Vol 2442(8)* **135–142**

[9] Dempster W T and Gaughran G R L 1955 Properties of body segments base on size and weight *The American Journal of Anatomy, Vol. 120(7414)* **33-54**

IOP Conf. Series: Materials Science and Engineering **931** (2020) 012022 doi:10.1088/1757-899X/931/1/012022

The Design of Rowing Ergometer Based on the Anthropometry of Acehnese Male Athletes

I Hasanuddin[1], Husaini[1], M N Rizki[1], Akhyar[1], M Nasruddin[2], N Amir[2], Y Marlina[2], L Asmawati[2]
[1]Mechaninal and Industrial Engineering Department, Universitas Syiah Kuala, Darussalam, Banda Aceh, Indonesia
[2]Physical Education Department, Universitas Syiah Kuala, Darussalam, Banda Aceh, Indonesia

E-mail: iskandarhasanuddin@unsyiah.ac.id

Abstract. A rowing athlete requires effective and systematic physical training. Thus, an ergometer is very influential on the endurance of an athlete's body and becomes a vehicle of simulating groundwater rowing. However, the ergometer tool currently used has no specifications and dimensions that are in accordance with the anthropometry of male rowing athletes in Aceh. Based on these conditions, the rowing ergometer tool will be designed based on the anthropometry of rowing athletes in Aceh. Data retrieval was done by taking samples from six male rowing athletes. The data were then tested for the level of trust, accuracy, uniformity, and adequacy. From the results of data analysis, the design of the ergometer tool was in accordance with the anthropometry of male rowing athletes in Aceh. This situation can build comfort for male athletes and increase the endurance of their muscles. This includes maintaining the tempo of rowing and being able to move oars with long distances in a short time.

1. Introduction

Rowing is an activity that is technically, physically demanding and requires a sequence of coordinated actions by utilizing most of the muscle groups in the body [1]. To achieve a rhythmic and harmonious movement, it needs to be supported by biomotoric quality, biometrics, psychology, and other supporting aspects [2].

Rowing is one of the sports that have developed in Indonesia, especially in Aceh. Acehnese male rowing athletes are trained to foster physical conditions for maximum achievement but have not found satisfactory results from the training and coaching [3]. The important exercise factors are technical, tactical, physical and mental factors. Of the four factors, the physical factor is the most important in all sports, so a planned and systematic training program is needed. Moreover, in systematic training, the athletes are trained to form pattern movement and angle while they row. This is so that the constant speed can be achieved in the rowing process [4].

From the explanation above, the author assumes that a rower athlete must have good endurance to be able to maintain the tempo of rowing. Considering the rowing sports must be carried out in rivers, lakes, or calm seas, it is certainly very risky if there is a strong current or rain. To overcome this, the rowers had to do training in the room using a simulation machine, which is commonly called the rowing ergometer machine.

Research has shown that the ergometer is the right training device, considering the value of a higher grip strength when rowing [5]. However, existing ergometer devices are not designed based on the shape and size of Indonesian athletes, especially Acehnese athletes from Aceh, one of the provinces in Indonesia. This becomes the foundation of conducting this research. In this study, the researchers designed a rowing ergometer tool kit taking into account the influence of the anthropometry of the Acehnese rowing athletes.

This study aims to design rowing ergometer devices and determine the effect of the Acehnese male athletes' anthropometry on the design of the ergometer devices. The output of this study is hoped to have the athletes be comfortable when using their sports' tools so that they increase the endurance of their muscles to maintain the tempo of rowing and move the oars in long distances in a short time.

2. Literature review

2.1. Rowing ergometer

The ergometer is designed by adapting rowing movements in water, but this activity is carried out on land. This tool serves to simulate the action of rowing in water and becomes a training tool that is fast and easy. Generally, the friction contact forces on the ergometer occur between flywheels that use air such as in Figure 1 (a) or water that moves the opposite as appeared in Figure 1 (b). The user pulls a rope that rotates the loading system with an intermediary tool i.e. pulleys, just like rowing in the water. Thus, the value of strength and speed on the pulley's spinning wheel is obtained by using a measuring instrument [6].

(a) (b)

Figure 1. (a) Air rowing ergometer (b) Water rowing ergometer [7].

The ergometer that is currently available still has some disadvantages, especially in its dimensions. The dimensions of the ergometer in the market today are not designed based on the anthropometry of the Asian population, especially Indonesians, which generally has smaller postures. Thus, users of rowing ergometers in Indonesia, especially in Aceh, get several obstacles during its usage.

2.2. Anthropometry

Anthropometry is a field of study about humans that deals with measurement, especially in measuring the body, shape, strength, and work capacity [8]. It is a collection of numerical data that deals with the physical characteristics of the human body, namely the size, shape, and strength, and application of the data for handling design problems [9]. Many studies have used anthropometry in designing sports tools and various equipment that are essential to humans' needs. Among the studies is the design of *sepak takraw* balls [10], exoskeleton to assist in manual lifting [11], headgear for soccer players [12], dynamometer for cutting forces [13] and even chairs [14].

Anthropometric measurements obtained from measurements are applied in terms of the design of rowing ergometer devices as discussed in this study. The results of studies in anthropometry can determine the shape, and the exact dimensions related to the product that will be designed according to the humans who will use the product [15].

2.3. Anthropometry dimension
According to Irdiastadi and Yassierli [16], the anthropometric database is the main source of information needed for design. There are 17 dimensions of user anthropometry that are closely related to determining the shape and dimensions of a rowing ergometer tool shown in Figure 2.

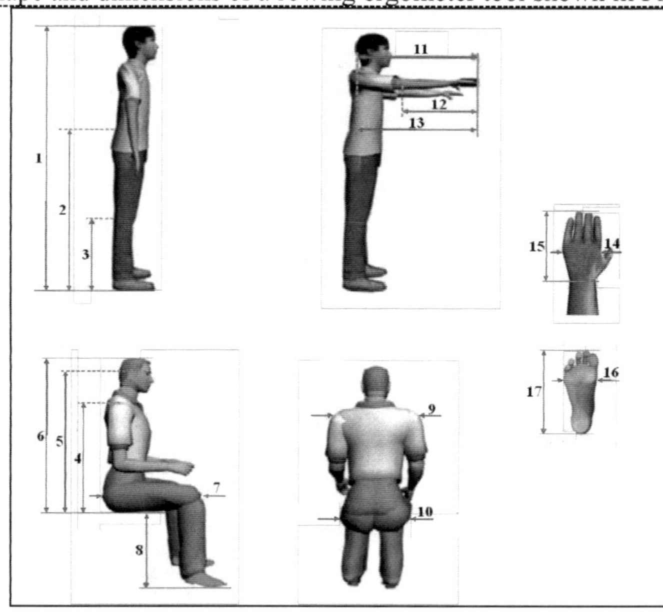

Figure 2. Anthropometric dimensions measured in the sample [17].

Meanwhile, Table 1 shows the 17 dimensions to be measured from a human body to obtain anthropometric data.

Table 1. Description of the dimensions of anthropometry.

No	Dimension	No	Dimension
1.	Upright body height	10.	Hip width
2.	High waist standing	11.	Future reach
3.	Knee height standing	12.	Long forearm
4.	Sitting shoulder height	13.	Long upper Arm
5.	Eye height sitting	14.	Hand width
6.	Sitting height	15.	Length of the palm
7.	Buttock knee	16.	Leg width
8.	Popliteal height	17.	Foot length
9.	Shoulder width	10.	Hip width

2.4. Determination of anthropometry and normal distribution data
Based on statistics, the size of humans in a population is around the average size, and a small portion of the sizes is on the two sides of the distribution. This implies that about 50% of the user population

IOP Conf. Series: Materials Science and Engineering **931** (2020) 012022 doi:10.1088/1757-899X/931/1/012022

can use the results of the design properly. While the remaining 50% cannot use the design properly; and this design is based on an average concept only [18].

Normal distribution methods are commonly used in establishing anthropometric data. Then, the percentile determination can be done with a normal distribution table. The 95-th percentile size anthropometry will show the "largest" body dimension and the 5-th percentile describes the "smallest" body dimension [19].

2.5. Data testing

To realize a product design that has flexibility and "capable of conformity" with a certain size range, it is necessary to obtain the right anthropometric data. Consequently, it is necessary to test anthropometric data to determine the correctness of this data. The following are the stages in testing anthropometric data:

a. Level of trust and level of accuracy

The level of trust displays the value of the belief of a researcher that the results of the calculations obtained to meet the requirements of accuracy. While the level of accuracy describes the highest deviation from the measurement results that are understood by researchers.

b. Data uniformity test

Test of data uniformity is a stage in statistics that aims to obtain a certain level of confidence in the data and shows that all are within the control limit. The data uniformity test serves to minimize the variety of existing data by removing extreme data or data that is outs of the control limits [20].

The steps in the data uniformity test are determining the average, standard deviation, upper control limit (UCL) and lower control limit (LCL).

To determine the average of the data set, Equation 1 is used

$$\bar{X} = \frac{\sum Xi}{n} \tag{1}$$

Where:
$\bar{X}$ = Average data from observations
Xi= Data on the measurement results of i
n = The number of numbers observed

In determining the standard deviation, Equation 2 is used.

$$\sigma = \sqrt{\frac{\sum(x_i - \bar{X})^2}{n-1}} \tag{2}$$

Where:
σ = Standard deviation
x_i = Results of measuring data

To get the UCL and LCL, Equations 3 and 4 are used.

$$UCL = \overline{X} + k(\sigma) \tag{3}$$

$$LCL = \overline{X} - k(\sigma) \tag{4}$$

Where:
k = Coefficient of confidence level

a. Test the adequacy of data

In this study, testing the adequacy of data was used to test the data taken as sufficient or not by knowing the value of N '. If the value of N'≤ N, then the data measurement result is said to be sufficient and no data collection is needed anymore. The following are the test steps for data sufficiency by using Equation (5)[20].

$$N' = \left[\frac{\frac{k}{s}\sqrt{N(\sum_{i=1}^{n} x_i^2) - (\sum_{i=1}^{n} x_i)^2}}{(\sum_{j}^{n} x_i)} \right]^2 \quad (5)$$

Where:
N' = Number of observations that must be done
x_i = Results of i-size data
s = The level of accuracy that should be
k = Coefficient of confidence level

2.6. Determining percentile values
The standard deviation and mean must be done first to determine the percentile value as the basis for the design of this ergometer device; 95-th for the large percentile size, 50-th for the average and 5-th percentile size for the small percentile size. Equation (6) is used to determine the dimension value of the specified percentile [20].

$$X = \bar{X} \pm Z\sigma \quad (6)$$

Where:
X = Value for the selected percentile
$\bar{X}$ = Average value
Z = Constants for the selected percentile
σ = Standard deviation

3. Methodology
Preliminary data collection for design considerations was carried out at the Krueng Lamnyong Edge Rowing Training Center, Banda Aceh, Indonesia.

3.1. Data collection design
In designing this rowing ergometer tool to obtain the required data, the researcher carried out anthropometric data collection on six Acehnese male rowing athletes. Males were chosen because, at the moment, only male athletes are available. This is by using the field research techniques that are done directly in the field to obtain anthropometric data from the samples by physical measurements.

3.2. Identification of athlete's anthropometry
In general, there are 36 measuring parameters of the human body. But in this study, only 17 parameters were used (see Table 3) because they have large influences on the design of this rowing ergometer. Anthropometric data processing uses the concept of percentiles.

Table 2. Anthropometric dimensions

Dimension	Information	Dimension	Information
D1	Upright body height	D10	Shoulder width
D2	High waist standing	D11	Hip width
D3	Knee height	D12	Knee-length
D4	Sitting height	D13	Popliteal height
D5	Eye height sitting	D14	Hand width
D6	Sitting shoulder height	D15	Hand length
D7	Long upper arm	D16	Leg length
D8	Long forearm	D17	Leg width
D9	Forward reach		

3.3. Data processing
The description of the research design process begins with the collection of anthropometric data for the Acehnese male athletes. Then, data processing was done by testing data to determine the level of accuracy and trust, data uniformity test, and test data adequacy. Afterward, determining the concept of percentiles so that the anthropometric data taken can be a reference in determining the concept of rowing ergometer tool design.

4. Results and discussion
4.1. Anthropometry data
Anthropometric data is needed so that the ergometer tool to be designed has a size that fits the users, in this case, the Acehnese rowing athlete. With 17 measurement criteria to be taken according to the design requirements as displayed in Table 4.

Table 3. Anthropometric measurement data for Acehnese male athletes.

No.	Dimensions measured	Measurement Subject					
		1	2	3	4	5	6
1.	Seated Height	93,5	93,8	86	84,5	92	84,3
2.	Eye Height Sitting	82,5	82,3	71,9	74,5	80,5	71,3
3.	Sitting Shoulder Height	61	62,8	59,2	60,5	61	58
4.	Long Upper Arm	37	41	35,7	39	37	34,9
5.	Long forearm	48,7	45,3	43,3	45,3	48,7	43
6.	Popliteal Height	45	45,9	46,8	44	45,8	47
7.	Buttock knee	58	58,7	53	53,6	57,8	63
8.	Shoulder Width	35,8	36	38,7	40,1	36,8	40
9.	Hip width	35,1	32,3	32	30,4	35,7	32
10.	Upright Body Height	174,1	174,5	163,8	164	173,6	156,1
11.	High Waist Stand	98	100,9	87	93	97	86
12.	Knee Height Standing	48,5	48,3	43,8	47	47	43
13.	Future reach	85,7	87,5	77	84,3	85,7	75
14.	Hand Width	8,8	9,5	9,5	8,7	8	9
15.	Length of the palm	18,5	20,4	18,2	19,4	17,8	17
16.	Foot Length	26,5	27,5	24,5	24,6	27,5	24
17.	Leg Width	7,1	7,8	7,5	7,5	7,1	7,6

4.2. Anthropometry data uniformity test
Calculation of data uniformity test in the design of this rowing ergometer tool uses the Excel program by using data obtained from direct measurements of the anthropometry of Acehnese male athletes (see Table 4). Processing is carried out repeatedly when there is data that passes the lower control limit (LCL) and upper control limit (UCL) when testing.

From the test results, uniform information from the overall uniformity test on 17 anthropometric dimensions of Acehnese male athletes are shown in Table 5.

Table 4. Anthropometric data uniformity test results.

No	Dimensions	Symbol	$\bar{X}$	σ	UCL	LCL	Description
1	Upright Body Height	TBT	167.68	6,383	180.45	154.92	equal
2	High Waist Stand	TPB	91.15	8.100	107.35	74.95	equal
3	Knee Height Standing	TLB	46.27	1.911	50.09	42.44	equal
4	Seated Height	TD	89.02	4.083	97.18	80.85	equal
5	Eye Height Sitting	TMD	77.17	4.600	86.37	67.97	equal
6	Sitting Shoulder Height	TBD	60.23	1.45	63.15	57.32	equal

7	Long Upper Arm	PLA	36.97	2.44	41.85	32.08	equal
8	Long forearm	PLB	45.07	2.57	50.22	39.91	equal
9	Future reach	JTK	82.03	5.02	92.08	71.99	equal
10	Shoulder Width	LB	36.40	3.20	42.80	30.,00	equal
11	Hip width	LP	32.08	2.21	36.51	27.66	equal
12	Buttock popliteal	PK	57.35	3.36	64.08	50.62	equal
No	Dimension	Symbol	$\bar{X}$	σ	BKA	BKB	equal
13	Popliteal Height	TP	44.92	1.97	48.86	40.97	equal
14	Hand Width	LT	8.95	0.88	10.72	7.18	equal
15	Length of the palm	PTT	18.62	0.98	20.59	16.64	equal
16	Foot Length	PTK	26.10	2.40	30.90	21.30	equal
17	Leg Width	LK	8.35	1.36	11,08	5.62	equal

4.3. Data adequacy test

The next stage of processing anthropometric data is testing the adequacy of data. In doing so, the confidence level used is 95% and the figure level of accuracy is 5%. Table 6 shows the results of testing adequacy.

Table 5. Adequacy of anthropometric data

No.	**Measurement**	**N**	**N'**	**Description (N'≤N)**
1	Upright Body Height	6	3	Adequate
2	High Waist Stand	6	6	Adequate
3	Knee Height Standing	6	4	Adequate
4	Seated Height	6	4	Adequate
5	Eye Height Sitting	6	6	Adequate
6	Sitting Shoulder Height	6	1	Adequate
7	Long Upper Arm	6	6	Adequate
8	Long forearm	6	4	Adequate
9	Future reach	6	5	Adequate
10	Shoulder Width	6	4	Adequate
11	Hip width	6	5	Adequate
12	Buttock popliteal	6	5	Adequate
13	Popliteal Height	6	1	Adequate
14	Hand Width	6	5	Adequate
15	Length of the palm	6	6	Adequate
16	Rigid Palm Length	6	5	Adequate
17	Leg Width	6	2	Adequate

4.4. Percentile calculation

In the design of this rowing ergometer device, a 95-th percentile was used for large percentile size, 50-th percentile for average percentile size, and 5-th percentile for small percentile size. Table 7 shows the percentile size used in data.

Table 6. Percentile size data.

No.	**Anthropometry**	**Dimension**	**Percentile (cm)**		
			5-th	**50-th**	**95-th**
1	Upright Body Height	D1	157.2	167.6	178.1
2	High Waist Stand	D2	85.4	91.1	101.8
3	Knee Height Standing	D7	43.1	46.2	49.4
4	Seated Height	D3	82.3	89	95.7
5	Eye Height Sitting	D4	69.6	77.1	84.7

6	Sitting Shoulder Height	D5	58.4	60.4	62.4
7	Long Upper Arm	D10	34.3	37.2	40.2
8	Long forearm	D11	41.6	45.5	49.4
9	Future reach	D12	75.3	82.5	89.6
10	Shoulder Width	D13	35.1	37.9	40.6
11	Hip width	D9	30.2	32.9	35.6
12	Buttock popliteal	D6	52.9	57.3	61.7
13	Popliteal Height	D8	44.3	45.7	47.1
14	Hand Width	D16	8.2	8.9	9.6
15	Length of the palm	D15	16.9	18.6	20.2
16	Rigid Palm Length	D17	23.4	25.7	28
17	Leg Width	D18	7	7.4	7.8

Table 7 shows the calculation obtained in the 5th, 50th, and 95th percentiles of each body dimension related to the design of the rowing ergometer tool. These values were the basis to determine the dimensions of the rowing ergometer tool for the Acehnese male athletes.

4.5. Determining the dimensions of the rowing ergometer tool

Based on calculations of the Acehnese male athletes' anthropometric data percentile, the next step was to determine the dimensions of the parts of this ergometer tool. The dimensions of the ergometer are presented in Table 8.

Table 7. The size of ergometer tool.

No.	Body size	Rowing Ergometer Tool Dimensions	Percentile	Design Size (cm)
1.	Upright Body Height	Tool Length	95	178.15
2.	High Waist Stand	Distance Between Leg Chairs with Chairs	95	101.82
3.	Knee Height Standing	Distance Between Seats with Leg Chairs	95	49.40
4.	Seated Height	Tool Height	95	95.71
5.	Eye Height Sitting	Height of Strength Indicator	5	69.62
6.	Sitting Shoulder Height	Grip Height	5	58.43
7.	Long Upper Arm	Distance to the user's body	5	34.31
8.	Long forearm	Distance to the user's body	5	41.67
9.	Future reach	Distance to the user's body	5	75.39
10.	Shoulder Width	Handle Length	95	40.69
11.	Hip width	Chair Length	95	35.62
12.	Buttock popliteal	Seat Width	50	57.35
13.	Popliteal Height	Seat Height	50	45.75
14.	Hand Width	Handle Size	95	9.60
15.	Length of the palm	Handle Size	5	16.99
16.	Foot Length	Leg Length	95	28.06
17.	Leg Width	Foot Mount Width	95	7.80

4.6. Design result

From the anthropometric data obtained, the dimensions of the ergometer tool were determined. By using the 2018 Autodesk Inventor Professional software, the design is finally constructed based on the anthropometric data of the Acehnese male athletes (see Figure 3). The size of the ergometer is based on the Acehnese anthropometry.

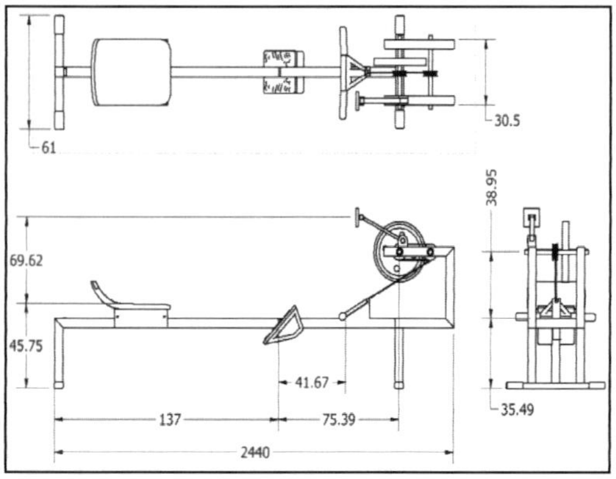

Figure 3. Rowing ergometer tool design based on the Acehnese athlete's anthropometry in form of Isometric view.

Figure 4. Rowing ergometer tool design based on the Acehnese athlete's anthropometry in form of three dimensional image.

5. Conclusion

From the implementation of the research and design, several conclusions are drawn:

1. There are 17 anthropometric dimensions of Acehnese male rowing athletes that influence the design of this ergometer tool. Anthropometric data were taken from a sample of six male rowing athletes who represented the population of rowing users.
2. For data analysis, testing the adequacy test, data uniformity, and determination of percentiles as the basis of the design were conducted. From the test results, the seventeen anthropometric dimensions revealed uniform data and can be said to be sufficient through testing the adequacy of data. The 95-th, 50-th, and 5-th percentiles were found as shown in Table 7.
3. The results of determining the percentile value were used to design the ergometer tool, as presented in Table 8. The product design of the rowing ergometer tool is displayed in Figure 3.
4. Despite the design of the rowing ergometer tool was accomplished in this research, some limitations were found during this research and are expected to be considered by future researchers. First, data were only taken from male athletes, and thus, if women rowing athletes decides to participate in this sport in the future, another research to obtain their anthropometric data should also be conducted to design a rowing ergometer tool suitable for the females. Second, this research used a manual loading system, hence, future research should consider using the automatic loading system to add loading variation on the athletes while collecting data.

6. Acknowledgement

In carrying out this research, the authors express their gratitude to the Work System Design and Ergonomics Laboratory, Construction and Design Laboratory, Computational Mechanics Laboratory, Design, and Manufacture Laboratory at Department of Mechanical and Industrial Engineering, Faculty of Engineering, Universitas Syiah Kuala for the facilities provided in supporting this research.

7. References

[1] Cerne T, Kamnik R, Munih M 2011 The measurement setup for real-time biomechanical analysis of rowing on an ergometer. *Measurement, Vol. 44* **1819-1827**

[2] Nurjaya D R 2002 *Instrumen pemanduan bakat "dayung" ["Rowing" talent scouting instruments]* Jakarta Direktorat Pelajar dan Mahasiswa

[3] Mizuar M 2011 Evaluasi kondisi fisik dominan atlet dayung Pengprov Aceh tahun 2011 [Evaluation of the dominant physical condition of Aceh Province rowing athlete in 2011] *Bachelor's thesis Universitas Syiah Kuala, Banda Aceh*

[4] Asmawati L, Amir N, Marlina Y, Hasanuddin I, Putra S 2019 The analysis of hand motion angle and distance of 200 meters rowing athletes administrators the Province of Aceh, *IOSR Journal of Humanities and Social Science (IOSR-JHSS) Vol. 24*(9) **20-25**

[5] Marcolina G, Lentolab A, Paolia A, Petronec N 2015 Rowing on a boat versus rowing on an ergo-meter: A biomechanical and electromyographical preliminary study. *Procedia Engineering, 112*: **461-466**

[6] McMurray R G, Ondrak K S 2007 Energy expenditure of athletes. In J. A. Driskell & I Wolinsky (Eds.), *Sports nutrition: Energy metabolism and exercise* **127-157**

[7] Fernandez-Ruiz B 2018 *Water rower vs. concept 2 power.* Retrieved on October 29, 2018, from http://www.olympum.com/sports/WaterRower-vs-Concept2-Power/

[8] Pheasant S, Haslegrave C M 2005 *Bodyspace: Anthropometry, ergonomics and the design of work, 3rd ed* Boca Raton: CRC Press

[9] Stevenson M G 1989 Lecture notes on the principles of ergonomics *Sydney: University of New South Wales*

[10] Hasanuddin I, Taha Z, Yusoff N, Ahmad N, Ghazilla R A R, Husni, Tuan Ya T M Y S 2015 Investigation of the head impact power of a *sepak takraw* ball on *sepak takraw* players. *Movement, Health & Movement Journal, Vol. 4(2)* **47-58**

[11] Salam R, Iqbal M, Hasanuddin I 2018 Desain dan analisis artificial exoskeleton pada prajurit TNI [Design and analysis of artificial exoskeleton in TNI soldiers]. *Jurnal Optimasi Sistem Industri, Vol. 17(2)* **135-142**

[12] Hassan M H A, Taha Z, Hasanuddin I, Mokhtarudin M JM 2018 *Analysis of protective headgear* Singapore Springer

[13] Yaldız S, Ünsaçar F 2006 A dynamometer design for measurement the cutting forces on turning. *Measurement, Vol. 39(1)* **80-89**

[14] Ghazilla R A R, Taha Z, Kamaruddin S, Hasanuddin I 2010 Pilot investigation on the mismatches of classroom furniture and student body dimensions in Malaysian secondary schools. *Journal of Social Sciences, 6(2)* **287-292**

[15] Wignjosoebroto S 2006 *Ergonomi, studi gerak dan waktu [Ergonomics, the study of motion and time]* Jakarta PT. Guna Widya

[16] Iridiastadi H, Yassierli 2016 *Ergonomi, suatu pengantar* [Ergonomics, an introduction]. Bandung PT Remaja Rosdakarya

[17] Antropometri Indonesia, *Dimensi antropometri*. Retrieved on March 18, 2019, from http://www.antropometriindonesia.org/index.php/detail/sub/3/4/0/dimensi_antropometri

[18] Sokhibi A 2017 Perancangan kursi ergonomis untuk memperbaiki posisi kerja pada proses packaging Jenang [Ergonomic chair design to improve work position in the Jenang packaging process]. *Jurnal Rekayasa Industri, Vol. 3(1)* **61-72**

[19] Ulfa R 2017 *Perancangan ulang kursi penumpang becak motor dengan metode antropometri (Studi kasus di Kota Banda Aceh)* [Redesign of a motorized pedicab passenger seat using the anthropometric method (A case study in Banda Aceh)] *Bachelor's thesis Universitas Syiah Kuala, Banda Aceh*

[20] Kempe H R 1996 *Kempe's engineer's yearbook, Volume 101* London Morgan Bros.

IOP Conf. Series: Materials Science and Engineering **931** (2020) 012008 doi:10.1088/1757-899X/931/1/012008

Mapping Upstream and Downstream Process in The Patchouli Oil Industry Using Supply Chain Operations Reference Model Version 12.0 (SCOR 12.0)

Hazful Maizi[1], Hasan Yudie Sastra[2] and Arhami[3]
[1] Industrial Engineering Master Program, Department of Mechanical Engineering and Industry, Universitas Syiah Kuala, Banda Aceh, Indonesia
[2,3] Department of Mechanical and Industrial Engineering, Faculty of Engineering, Universitas Syiah Kuala, Banda Aceh, Indonesia

E-mail: hazful@gmail.com

Abstract. Supply chain characteristics in the agricultural industry differ from the industry in general, the structure is more complex and unique according to its agricultural commodities. All of the aspects starting from the upstream are very important to consider because of their dependence on the suitability of the environment and the processes that are carried out on the pre-source. Patchouli oil industry which is mainly located in Aceh Jaya Regency is the object of this study due to the fact it is the largest supplier of patchouli oil in Aceh Province. It is estimated that patchouli oil which is one of the superior local wisdom product is becoming an interesting subject again. This is indicated by several local products that use it as the main raw material. Certainly, this will be difficult to be applied if the supply chain from the upstream to the downstream is not evaluated and upgraded. For this reason, it is necessary to do a mapping process on the vulnerable side, including *pre-source*, *source*, *make*, and *deliver*. This mapping used the SCOR method version 12.0 and adapted to the make-to-stock industry scheme. The output of this study is a reliable and measurable process system in the patchouli oil industry.

1. Introduction

The quality of Aceh Patchouli Oil has long been recognized internationally. The best Aceh Patchouli variety was Lhokseumawe Variety, Tapak Tuan Variety, and Sidikalang Variety. The varieties have become the priority crops for the cultivation in Indonesia. However, the patchouli oil deterioration and uncertainty price indicates the failure in the supply chain system. The supply chain system greatly determines its performance and reliability. Part of the system includes *source*, *make*, and *deliver*. This problem needs to be investigated and evaluated as an effort to restore the glory of the patchouli oil industry in Aceh.

This study used the Supply Chain Operations Reference Model version 12.0 (SCOR 12.0) approach method. The method is used to map and evaluate the processes that occur in supply chain system.

Literature Review

Aceh Patchouli Plants (*Pogostemon cablin* Benth). Patchouli is a type of shrub plant with the height up to one meter. Aceh Patchouli plants has the best quality of oil content and patchouli alcohol (PA). The standard quality of Aceh patchouli oil content is about 2.5%-5.0% [1]. Patchouli oil is fixative as a binding agent for other essential oil compounds. There is no alternative product that can replace the function and the role of the patchouli oil. The development of Aceh Patchouli has created the best variety called Tapak Tuan Variety, Lhokseumawe Variety and Sidikalang Variety. The advantage of Tapak Tuan Variety is the high level of patchouli alcohol (PA). Meanwhile the Lhokseumawe Variety has high oil content level. The Sidikalang Variety has resistance to bacterial wilt disease and nematodes. National Standardization Organization of Indonesia (SNI) has decided the standard quality of patchouli oil. The standard quality listed in SNI 06-2385-2006.

Table 1. The standard quality of patchouli oil.

No	Parameter	Requirement	
1	Color	Light yellow – Reddish brown	
2	Density 20 °C/20 °C	0.950 – 0.975	
3	Refractive index [nD20]	1.507 – 1.515	
4	Solubility in ethanol 90% at temperature 20 °C	Clear solvent/opalescence in comparison 1:10	
5	Acid number	Maximum 8	
6	Ester number	Maximum 20	
7	Optical rotation	(-)48° – (-)65°	
8	Content of Fe [mg/kg]	Maximum 25	
9	Patchouli chromatography profile using a detector	Gas Liquid Chromatography (GLC) and FID	
10	Component:	Minimum [%]	Maximum [%]
	α-Copaene	-	0.5
	Patchouli alcohol	30	-

Source: National Standardization Organization (BSN) ICS 71.100.60.

Supply Chain Management (SCM). Supply chain is a process of material flow, information flow, and payment flow in the trading system of the product starting from the upstream to the downstream aimed at the end user. Supply chain includes planning and controlling production, material handling, purchasing and payment, warehousing, distributing and shipping. The main purpose of the supply chain is to balance supply and demand. Supply Chain Management (SCM) is a concept in managing a supply chain activity to maximize customer value. SCM perspective is integrated services, which includes interconnection and relation between companies (linked) not only covered the first tier but also reached the second tier. The relationship includes the downstream and the upstream industry.

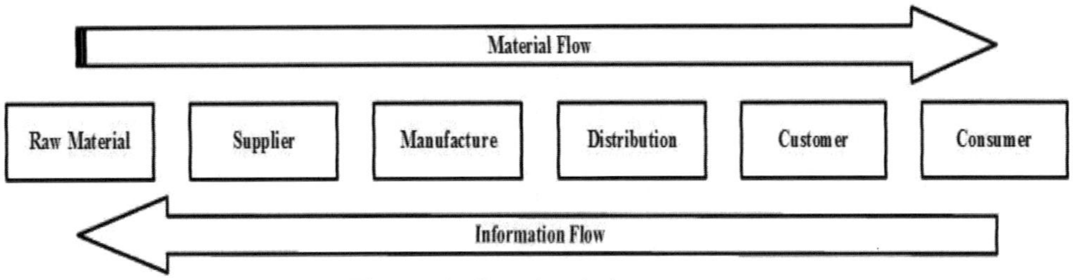

Figure 1. Supply chain system.

Supply Chain Operations Reference Model (SCOR 12.0). The SCOR model is a model that describe business activities oriented towards customer satisfaction. Supply Chain Operations Reference Model (SCOR 12.0) presents a unique business framework, work indicators, best practices and technology to support communication and collaboration between partners or entities within a supply chain. The model can improve the effectiveness of supply chain management. SCOR model forms as hierarchy with several levels of process. The first level is the type of process used to identify the scope of a supply

IOP Conf. Series: Materials Science and Engineering **931** (2020) 012008 doi:10.1088/1757-899X/931/1/012008

chain. The second level is the stage of formulating process categories that aim to configure supply chain. Meanwhile, the third level is a process that shows the elements of the process, determines the supply chain includes input/output, indicators of the process, and best practices can be implemented in the supply chain [3].

Table 2. Type and category process of SCOR

		SCOR Process					
		Plan	Source	Make	Deliver	Return	
Type of Process	Planning	P1	P2	P3	P4	P5	Category of Process
	Execution		S1-S3	M1-M3	D1-D4	SR1-SR3 DR1-DR3	
	Enable	EP	ES	EM	ED	ER	

2. Methodology

Procedure. The procedures performed in this study can be represented in the following flowchart.

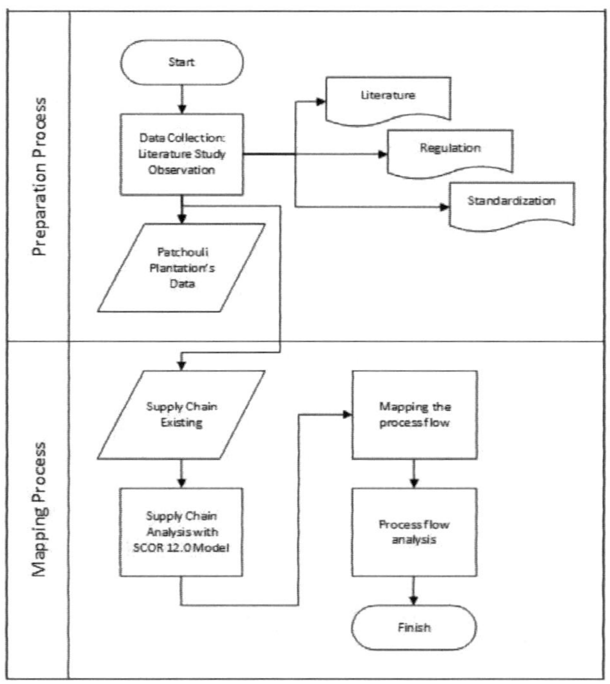

Figure 2. Preparation and mapping process.

Variable. Process categories that mapping with SCOR Model 12.0 are particularly processed in the supply chain system. The process categories are represented in the following table.

Table 3. Process area of SCOR Model.

Process	Process Area
PLAN	Supply and demand planning management.

SOURCE	Raw material inventory management, acceptance management, supplier selection management, procurement strategy, and performance management.
MAKE	Order management, production schedule, production activity, and engineering/customization.
DELIVER	Finished goods inventory management, supply and demand information management, product distribution, and installation management.
RETURN	Management of returning activities to suppliers or receiving returns from consumers.
ENABLE	Planning and execution arrangements.

Mapping Analysis

Supply Chain Existing System Configuration. Supply chain configuration that has been implemented includes several entities that play an important role in the patchouli oil supply chain system in Aceh Jaya Regency. The entities are Farmer I, Farmer II, Maker, Local Collector, The Cooperative, Local Customer, Exporter, and International Customer.

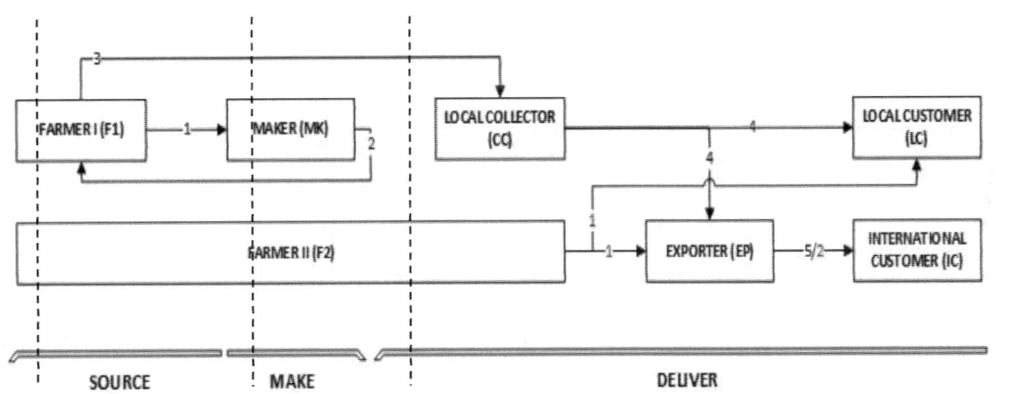

Figure 3. The entities of the existing supply chain system.

Type of Process	Pre-Source	Source	Make	Deliver
SC Actors	Farmer I Farmer II	Farmer I Farmer II	Farmer II Maker	Farmer II Local Collector
Main Activities	Cultivating Harvesting	Drying Chopping	Distillation Oil Separation Purification Packaging	Marketing
Output	Leaves Stems Roots	Raw Material	Patchouli Oil Waste	

Figure 4. Process flow of the supply chain system.

IOP Conf. Series: Materials Science and Engineering **931** (2020) 012008 doi:10.1088/1757-899X/931/1/012008

The following is a detailed mapping process.

- Plantation land management is not complete at all. The farmers use the available plantation land without land management first.
- The cultivation process is not carried out.
- Farmer plants the crop cutting directly on the land. The treatment has a high failure rate.
- In the process of maintaining the plants, not only weeding and pruning are not carried out, but also there is no effort to control pests, viruses, and diseases attack.
- Harvesting process is done by fully cutting to the base of the stem. As the result of this treatment, there is no secondary harvesting period.
- Harvested plants are placed on the ground, so that they are exposed to sunlight in relatively long period.
- Distribution activities include distribution of raw materials to the warehouse and distribution of the product to the customer. Generally, the distribution activities have not been carried out regularly and systematically.
- There is no representative warehouse.
- There is no selection of raw material quality.
- There is no fermentation process as an effort to improve the quality of patchouli oil.
- There is no control over the quality of production facility and infrastructure.
- There is no waste management.

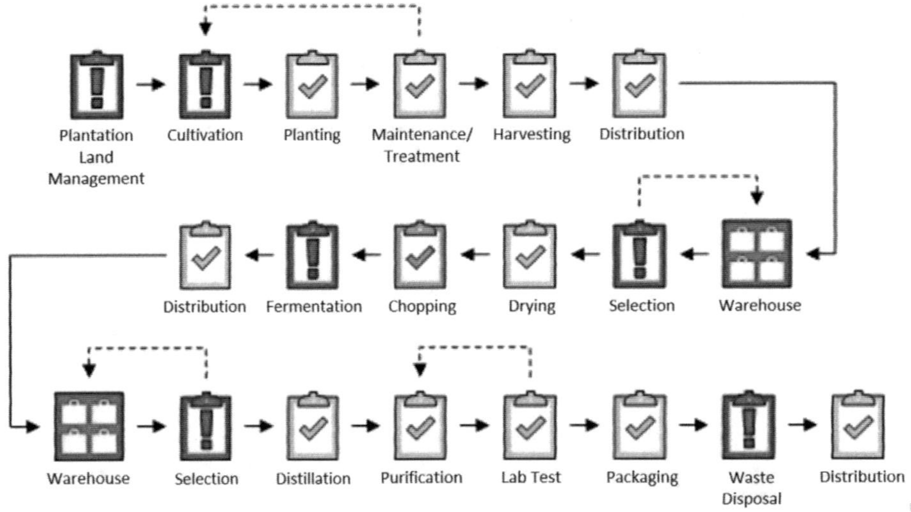

Fig. 5. Detail processes of the patchouli oil industry.

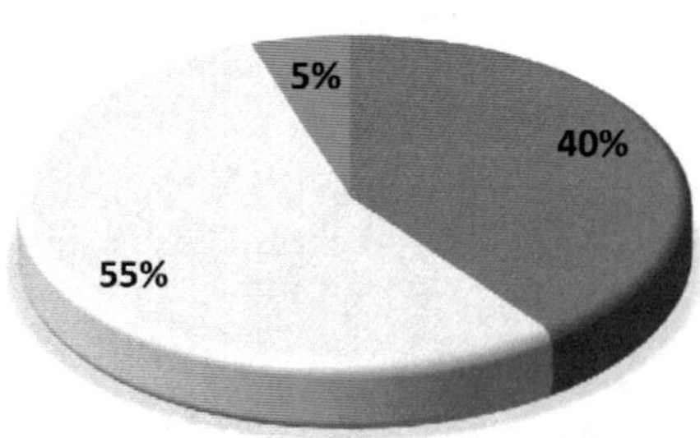

Figure 6. Percentage of detail processes.

Mapping of The Industrial Process Schemes with SCOR 12.0. The purpose of adjusting the process scheme with the SCOR Model version 12.0 approach is to identify the weaknesses of the supply chain system. It creates a strategy or an alternative improvement to the supply chain system. Eventually, the supply chain system will be more effective and reliable.

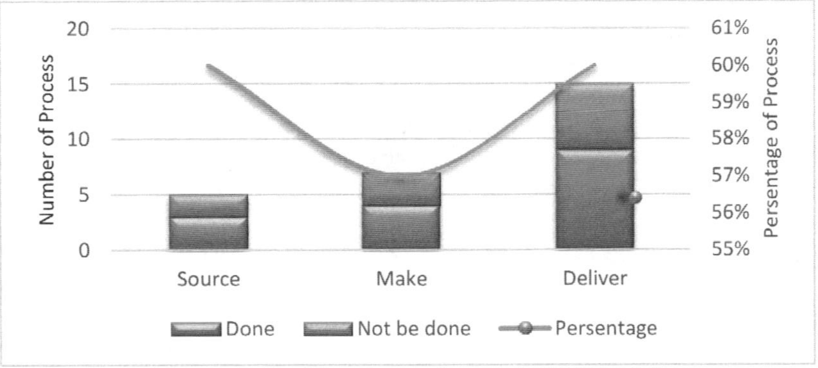

Figure 7. Profile of mapping with SCOR 12.0.

sS-Source Process. Mapping the *source* process metric in the patchouli oil industry only covers the source stocked product scheme, which is the process of raw material supply for production to maintain stability and availability of the products in inventory. The product is still available as a penetration of customer needs. The following is steps that fit the SCOR model version 12.0 for the *source* process.

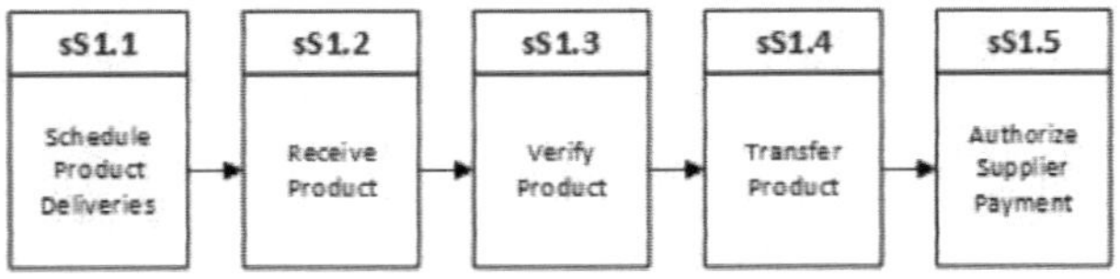

Figure 8. *Source* process scheme SCOR 12.0.

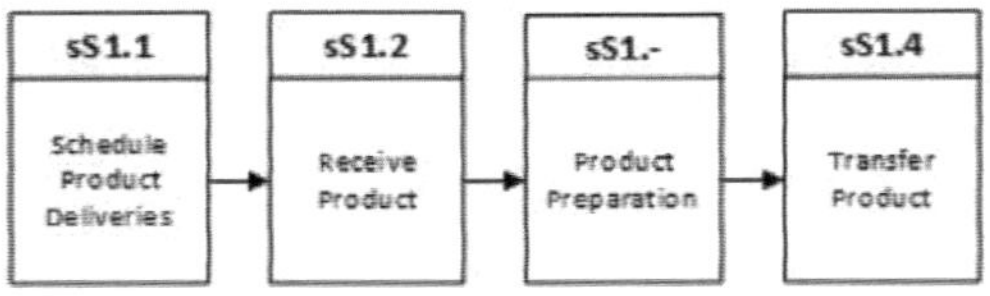

Figure 9. *Source* process scheme existing.

sM-Make Process. In the production process, most of the processes have been carried out in accordance with the process scheme approach in the SCOR Model. The mapped process scheme is make-to-stocked scheme.

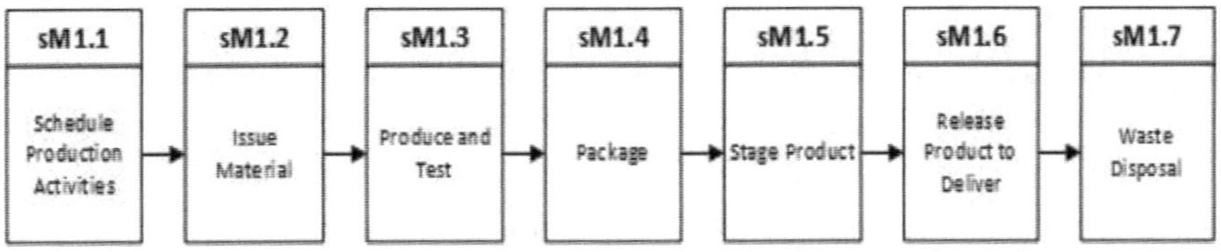

Figure 10. *Make* process scheme SCOR 12.0.

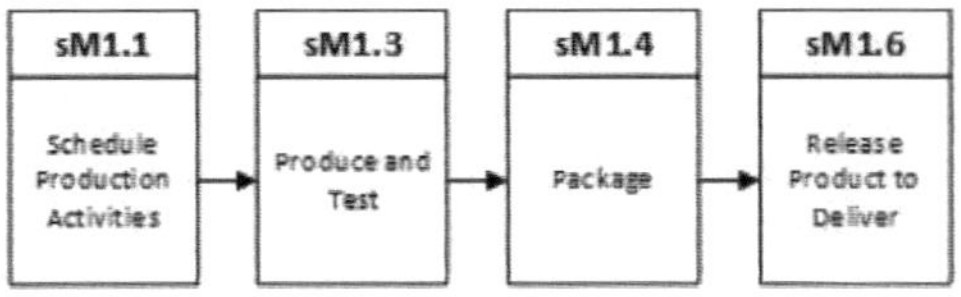

Figure 11. *Make* process scheme existing.

sD-Deliver Process. The *deliver* process is a process that determines the level of customer satisfaction with the accuracy of both product performance and delivery service. Thus it is very important to pay attention to every detail of the stages in the distribution process with the reference to the SCOR 12.0 approach.

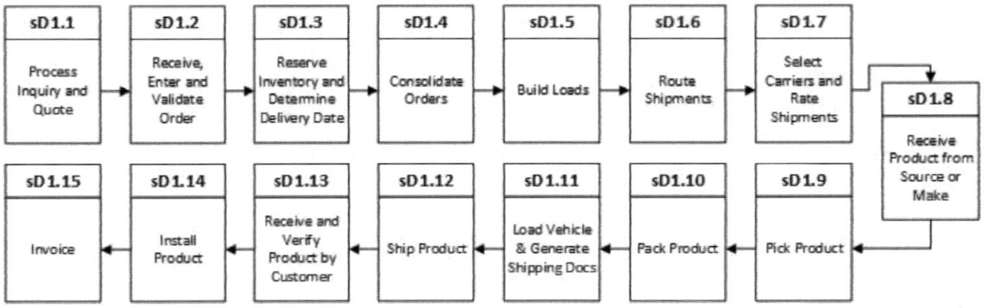

Figure 12. *Deliver* process scheme SCOR 12.0.

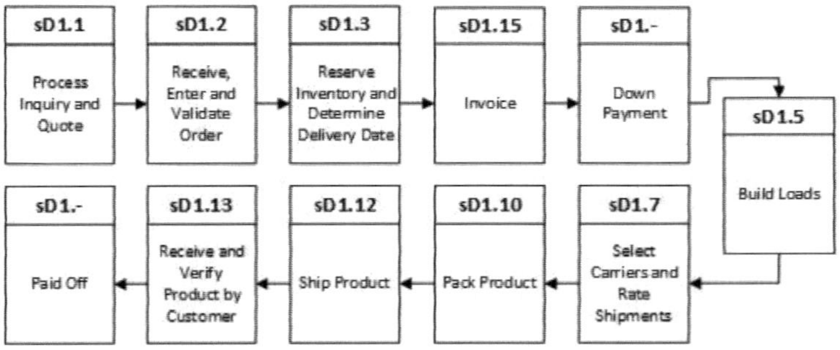

Figure 13. *Deliver* process scheme existing.

3. Summary

Mapping the supply chain model using the Supply Chain Operations Reference Model (SCOR 12.0) which includes *pre-source*, *source*, *make*, and *deliver* is very effective to identify the problems that occur in the supply chain. The calculation shows that only 5% of the process is carried out with systematic and measurable standardization process. Meanwhile about 55% of the process have not been completely conducted by following the standard and 40% of the process is not at all carried out by industry in accordance with the standard.

From the supply chain mapping using SCOR Model version 12.0, it was concluded that the average of the process that was not carried out on the existing supply chain system was 41%.

IOP Conf. Series: Materials Science and Engineering **931** (2020) 012008 doi:10.1088/1757-899X/931/1/012008

References

[1] Adharini, Dwi Wulan, 2009, *Aceh Patchouli Cultivation and Distillation (Pogostemon cablin Benth) at Deni Nursery and Gardening*, Sebelas Maret University, Surakarta

[2] Nasional Standardization Organization, 2006, *Patchouli Oil*, ICS 71.100.60, BSN, Jakarta, Indonesia.

[3] Paul, John, 2014, *Supply Chain Transformation with the SCOR Model*, PPM, Indonesia.

[4] APICS, 2017. *Quick Reference Guide Supply Chain Operations Reference (SCOR) Version 12.0*, APICS.

[5] Muharam, Salih, et all, *Quality Improvement of Patchouli Oil (Pogostemon cablin Benth) using Combination Method of Fermentation, Delignification and Distillation*, Jurnal Kimia Valensi Vol. 3 No. 2 (2017) 116-121.

[6] Wahyuniardi, Rizki, et all, *Performance Measurement of Supply Chain using Supply Chain Operations Reference (SCOR)*, Jurnal Ilmiah Teknik Industri Vol. 16 No. 2 (2017) 123-132.

[7] Mutakin, Anas & Musa Hubeis, *Management Performance Measurement of Supply Chain using SCOR Model 9.0*, Jurnal Manajemen dan Organisasi Vol. II No. 3 (2011) 89-103.

IOP Conf. Series: Materials Science and Engineering **931** (2020) 012015 doi:10.1088/1757-899X/931/1/012015

The planning of iron sand inventory with System Dynamic Approach

P D Sentia, Andriansyah, R D Lufika, M J Satria
Industrial Engineering Department, Universitas Syiah Kuala, Darussalam, Banda Aceh, Indonesia

E-mail: primadennysentia@unsyiah.ac.id

Abstract. The need for iron sand as a supporting material of the cement making is increasing, so companies require planning of inventory control of iron ore to get all the needs in terms of quality and quantity. Inventory control planning will be useful to meet the demand and reduce inventory wastage. In the production process, companies often experience a shortage of materials when demand soared and raw materials excessed when demand decreases. This study aims to determine the best ordering method next year so that the company does not experience the stock shortage. The system dynamics method was applied with the help of Powersim software in terms of modelling and simulating the iron ore ordering system in the form of several scenarios. Based on the result, the best scenario was to order the iron ore as much as 4,013 tons/month, which is twice the amount in the previous policy, and the average inventory of the iron ore was 10,943 tons/month.

1. Introduction

The rapid development of knowledge and technology enables improvement in the industrial sector, such as the iron sand industry. As the natural resources, the cement industry utilizes the iron sand as one of the components in the final products. It is necessary for a company to plan the iron sand raw materials control in terms of quality and quantity because of its high demand as the additional materials for the cement products. Inventory in terms of industry, it is necessary to pay attention to the idle resources for the smooth flow of the production process [1].

The case study is conducted at one of the iron sand industries located at Krueng Raya, Aceh Besar. The company applies the principle of sustainable iron sand in production to meet the fluctuation of the demand. The company faces the problem of inventory shortage when the demand increases and the inventory excess when the demand decreases. The reason for such a situation to occur in the company is because of the inaccurate demand forecast and inappropriate ordering schedule. The fluctuation of the demand shows the figure of customer demand, which is within the range of 2,393 tons to 4,814 tons. In order to solve this problem, a company has to supply the right amount of raw materials. One approach that will assist the company in planning the raw materials is the system dynamics approach, which shows a promising solution for the complex systems. This approach aims to understand the structural and dynamical portion of the complex system and to use that model to lead an effective and subtle organization [2].

There are numerous research studies in the area of inventories and system dynamics, such as to compare inventory models in an automotive [3], to determine safety stock [4], to reduce total inventory cost [5] and to improve complex inventory management [6]. Simulation is the practical or application of the model building step that represents the real system, or the future forecast, or the model experiment by itself, which aims to understand the system, to increase performance, or to design a new system with a designated dimension [7-8]. The predominance of the system dynamics is that the simulation will reduce total cost, time, and energy while preserving nature from the process of trial and error. The simulation process will be using the Powersim software for next year's long inventory planning.

The formulation of the problem in this study is to discover the variables that influence the system behaviour of the ordering numbers and to plan the raw materials with the system dynamics approach. Nevertheless, the simulation of a system dynamics will be showing the behaviour of the system model in terms of both the ordering numbers and the raw materials. The result of the simulation is the best scenario out of several different scenarios of the raw materials ordering quantity to satisfy the customer's demand.

The goal of this study is to observe the behaviour of the system in terms of the ordering numbers and the raw materials with Causal Loop Diagram (CLD) model and to decide the behaviour of the system in terms of the company's ordering quantity and inventory with Stock Flow Diagram (SFD) model. The result gave several scenarios of the raw materials ordering quantity that will comply with the demand from the customers. Therefore, this study consists of the modelling and the system dynamics simulation for obtaining the best scenario of inventory planning and raw materials ordering to avoid the inventory shortage situation.

2. Method

The simulation process includes several steps to solve the complexity of the models, i.e. [9]: formulation of the problems, building the models, data acquisition, transferring the models and formulations, verification, validation, the tactical and strategic planning, experiment, result in analysis, and implementation and documentation.

Firstly, it is necessary to identify the problems at the company to find the object of the study. Finding the object of the study, then the next step is further study regarding the solutions to the problems. The next step is to collect the appropriate data with the problems. There were two different data collection methods in this study, namely the primary and the secondary data. The primary data was the result of brainstorming with an expert to discover the problematic variable, which would be used as the structural system of the models for the CLD and SFD. The utility of secondary data was to develop the SFD models for simulation according to the problem in the company. The data included the iron sand demand, the iron sand inventory, and the utility of iron ore.

The next step was data processing. Data processing software called Powersim was used to process the data. There are three different parts of the data processing, i.e. CLD, SFD, and scenario simulation. There are three steps for processing the CLD. The first one is the preliminary CLD that is using the model from previous research. The second one is the model development into working CLD based on the problems at the company and the knowledge from experts. Before the final stage, it is necessary to brainstorm the model with experts. Finally, the final CLD is ready for the research [10].

For the development of SFD, it is necessary to transform the qualitative translation of CFD to a quantitative one. There are four steps for developing the SFD model [9]. The first step was to build the model structure which connected one component to another using Powersim software. The structure was developed using the logic of the problem; however, to strengthen the research, the development was based on previous research. The next step was inserting the value and formulation for the component of the model. This was the step of transforming the qualitative to quantitative. Verification was necessary to observe the unit consistency of the model. The last stage was to validate the SFD model with the behaviour validity test. Finally, if the model was valid, the scenario simulation could be performed. The best scenario for ordering and inventory management was the scenario that satisfied the overall demand from the customers and minimized the stock shortage in the company.

3. Result and discussion

There are three steps in developing the Causal Loop Diagram (CLD). The first step is to build the preliminary CLD based on the previous research on inventory problems. The second step is to add variables which are the result of brainstorming with stakeholder into the initial CLD for making the working CLD. The stakeholder then will conduct the verification and validation of the model as well as the literature review before finalizing the CLD model [10].

3.1. Preliminary CLD

The model in this research refers to the Business as Usual (BAU) inventory model from previous research by [11]. It is appropriate to use the previous BAU on developing the preliminary CLD. Only the appropriate variables on BAU will be the reference for developing the preliminary CLD; therefore, the selected variables and its connection were appropriate for this research, as presented in Figure 1, as the preliminary CLD. Loop B1 shows iron ore ordering, Loop B2 for iron ore inventory, while Loop B3 for production.

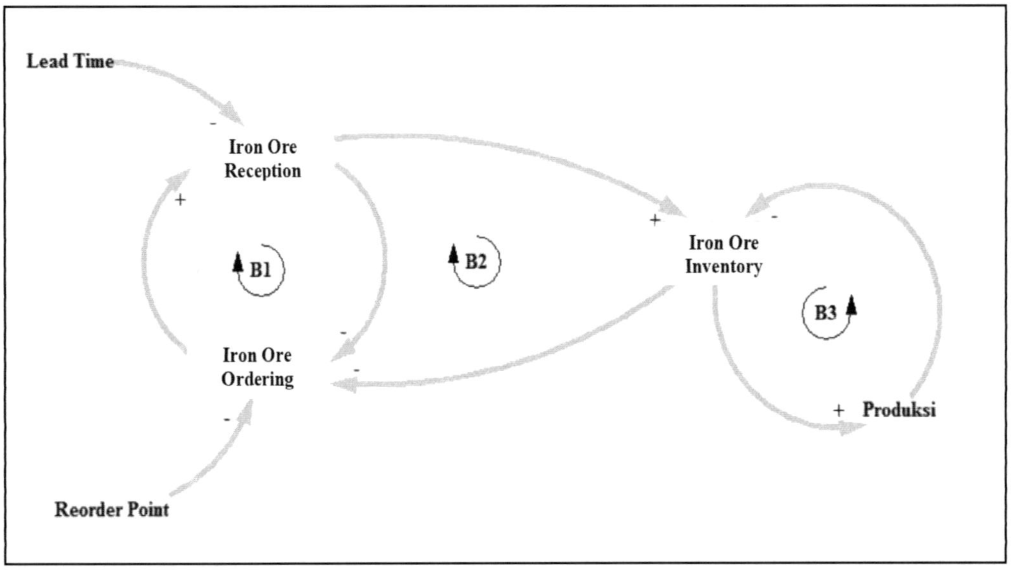

Figure 1. Preliminary CLD

3.2. Working CLD

The variables on the preliminary CLD was not adequate, therefore, the model required an additional variable for the conceptual model related to the planning of ordering and supplying raw materials. Figure 2 shows the working CLD, while Loop R1 in figure shows production enhancement. Figure 2 shows that the additional variables were related to the production of iron sand. The variables influenced the process of ordering iron sand and the inventory of iron ore. After that, with the help of stakeholder and literature review, the working CLD passed the verification process of structural and relation with the final CLD as a result

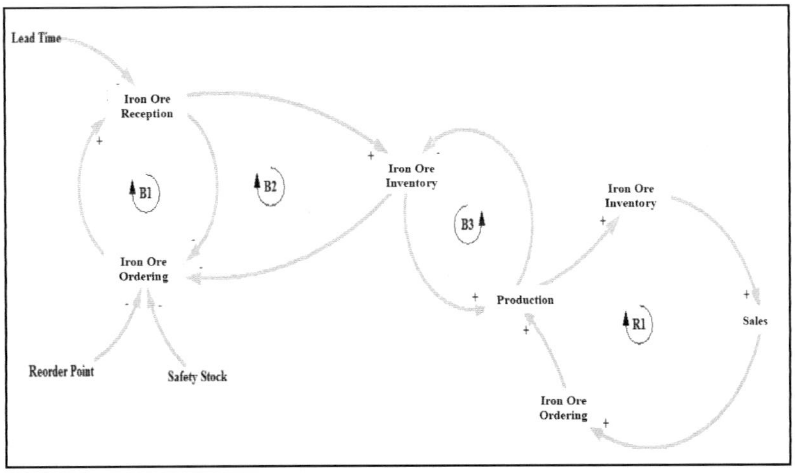

Figure 2. Working CLD

3.3. Final CLD
It is necessary to understand the structure of the modelling system for the accuracy level of the actual model, which represents the real system. The discussion with experts and literature review is necessary to achieve this goal. Two experts contribute their knowledge to build the model in this research.

The final CLD (figure 3) shows that there were two other variables which influenced the ordering of iron ore besides the iron ore inventory and the iron ore reception. The other variables were the re-order point and safety stock. The final CLD consisted of four different loops, and three of which were balancing loops and the other was a reinforced loop.

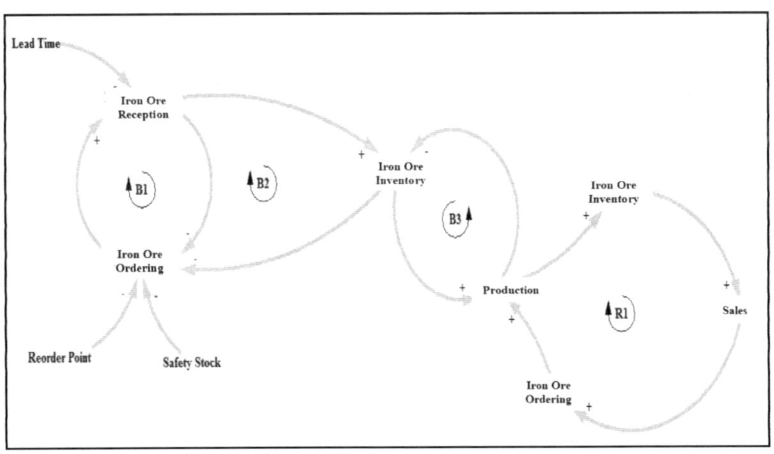

Figure 3. Final CLD

3.4. The SFD construction
The stock and flow diagram is more complicated than the causal loop diagram. It is necessary to have several sub-models to simplify the process of understanding and building the model. There were two sub-models in this research, i.e. the sub-model of the production process (Fig. 4) and the sub-model of ordering raw materials (Fig. 5). The final model required a combination of the two sub-models. The problem logic in building the SFD model refers to [12].

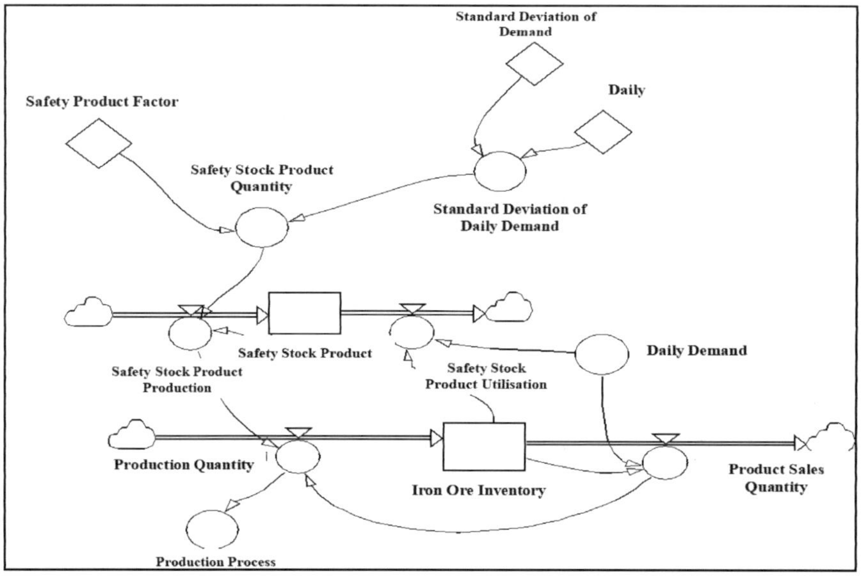

Figure 4. Sub-model of production process

There were several modifications to the model structure, variables relations, and variable constraints to build the SFD of the production process sub-models. The whole process of building the primary model of the production process was based on previous research with the same production process concept. The variables of production and sale quantity are the two main variables in determining the production process.

On the other hand, the building process of the raw materials ordering sub-model refers to the past research, the general logic of raw materials order, and the actual condition of the company. There were four significant variables which influenced the iron ore order process, i.e. the iron ore inventory level which relates to the iron ore reception, the iron ore utilization which was converted to the daily unit, the raw material in order which is part of the iron ore ordering and reception process, and the stock of first and second supplier which means the capacity of supplying the iron ore to the company.

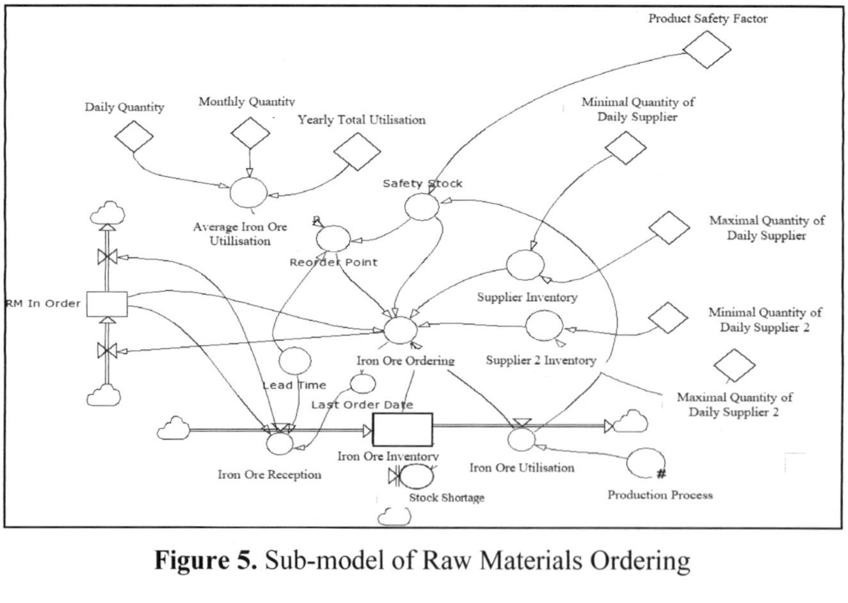

Figure 5. Sub-model of Raw Materials Ordering

Figure 6 shows the combination of the production process sub-model and the raw materials ordering sub-model. This is the SFD main model in this research.

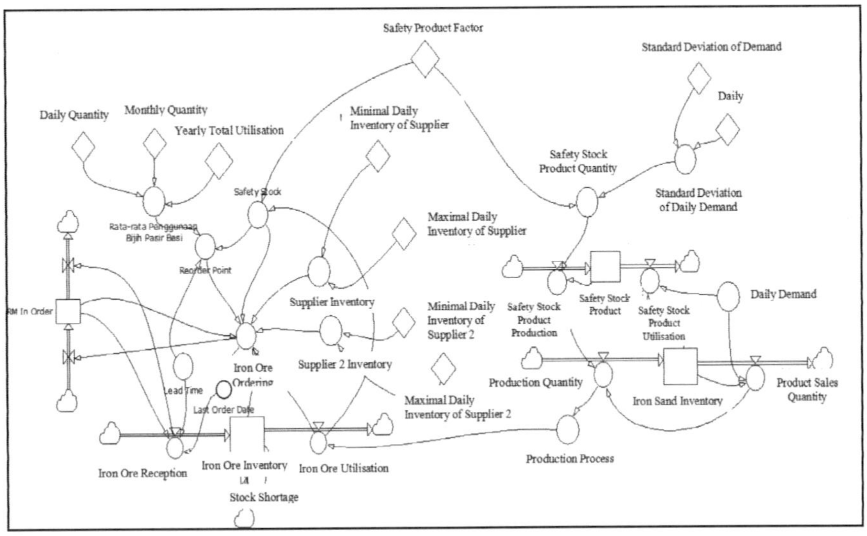

Figure 6. SFD main model

3.5. Simulation of the existing condition

Based on the SFD model, the simulation started with the existing condition and the alternative scenarios has not yet been considered. Figure 7 shows the time graph as the simulation result of the real situation. Figure 8 shows that the company will always experience the stock shortage for the next several years. This situation occurs because of the higher number of iron ore demand than the iron ore production quantity.

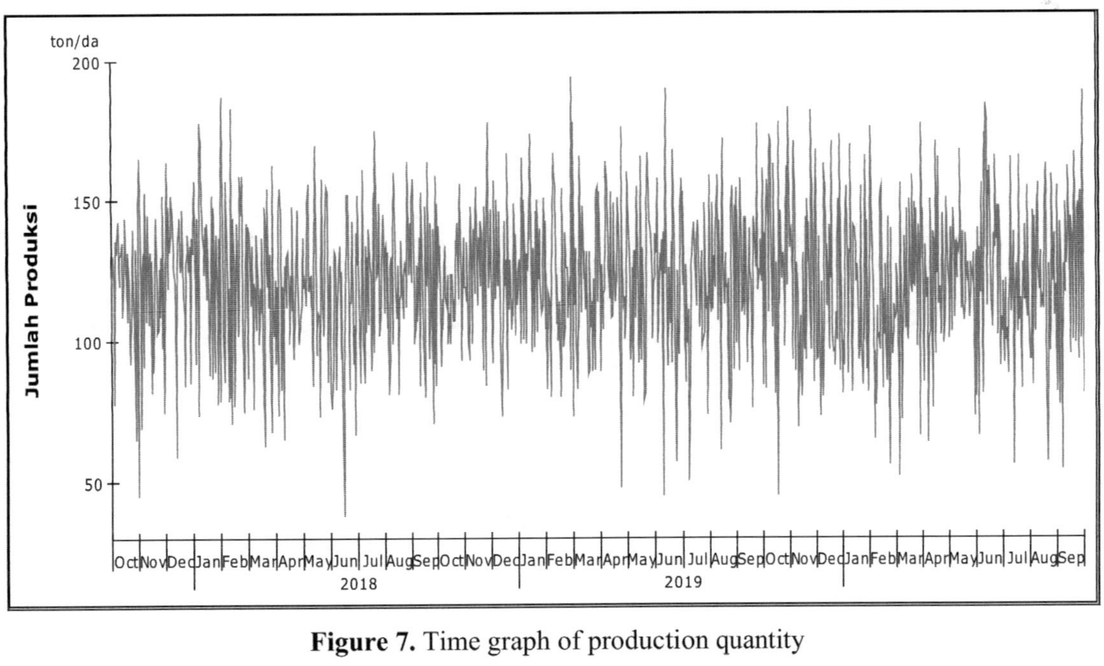

Figure 7. Time graph of production quantity

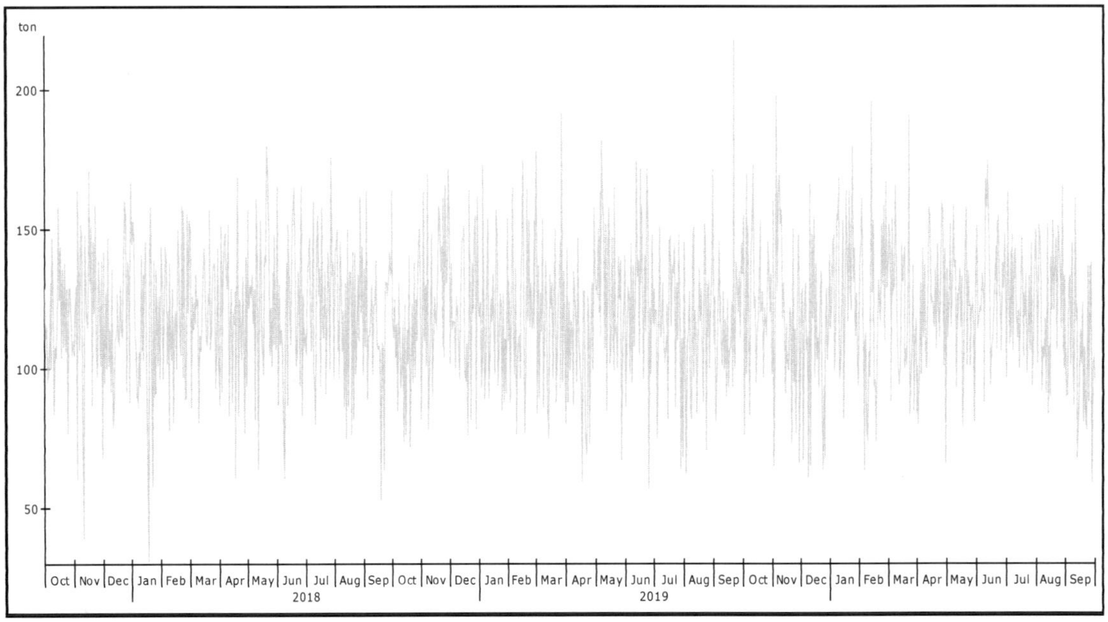

Figure 8. Time graph of iron ore utilisation

3.6. Verification

The CLD was verified after the theoretical verification and brainstorming with the company staff members. The reason is that, in the real system, the higher reception of iron ore equal to the higher inventory of iron ore. Therefore, the theoretical verification shows that there was no difference between the real situation and the CLD model in terms of the relationship between iron ore reception and inventory. On the other hand, the process of the SFD model verification requires the simulation test without demand, and the initial inventory was 350 tons.

In this research, the daily demand, the quotidian demand result, iron ore utilization, and iron ore production quantity were zero ton; and the inventory was the same as the initial stock, i.e. 350 tons. The test result shows that there was no obstacle in the simulation, which means that the SFD model was verified.

3.7. Validation

The validation process is necessary to find out whether the model simulation is valid with no error, and it is for comparing the structural model behaviour with the real situation. After the validation process, then the model is appropriate to represent the situation in the company. In this research, the validation process was performed using the behaviour validity test, which was based on historical data. Table 1 shows the historical data test result.

The historical data test by comparing the actual iron ore utilization with the simulation one, the mean error was 0.075656 or 7%. The result was lower than 10% with the level of accuracy requirements of 90%. If the validation value is under 10%, with 90% the level of accuracy, then the data is valid.

IOP Conf. Series: Materials Science and Engineering **931** (2020) 012015 doi:10.1088/1757-899X/931/1/012015

Table 1. Historical data test result

Month	Simulation (tons)	Actual (tons)	Errors
January	4272	4150	0.029398
February	4136	3226	0.282083
March	4099	3980	0.029899
April	4025	3587	0.122108
May	4116	3789	0.086302
June	4321	5079	0.149242
July	4293	4197	0.022873
August	4256	4244	0.002828
September	3765	4009	0.060863
October	3702	3570	0.036975
November	3894	4167	0.065515
December	4063	4145	0.019783
	Mean		0.075656

3.8. Replication Adequacy Test

It is necessary to have the replication adequacy test of the simulation result identification for running the simulation itself. Table 2 shows the replication adequacy test result in the simulation.

Table 2. Replication Adequacy Test Result

Month	Iron Ore Utilisation Simulation (tons)
January	4272
February	4136
March	4099
April	4025
May	4116
June	4321
July	4293
August	4256
September	3765
October	3702
November	3894
December	4063
Mean	4078
Standard Deviation	203,62

$$\text{Half width} = \frac{(2{,}26\ x\ 203{,}62)}{\sqrt{10}} = 145{,}52 \quad (1)$$

$$n' = \left(\frac{1{,}96\ x\ 203{,}62}{145{,}52}\right)^2 = (2{,}742)^2 = 7{,}521 \approx 8 \text{ replication times} \quad (2)$$

The calculation result shows that it requires at least eight replication times of the data replication test for the valid simulation result. This research consisted of 10 replication times, so the replication time for simulation was adequate. After the replication adequacy test, the next step was to simulate with alternative scenarios to get the best scenario for raw materials ordering policy that satisfies the iron sand demand.

3.9. Scenario simulation

The model aims to determine the raw material ordering quantity, which is more than the mean utilization of iron ore. There were three scenarios in this research, i.e. the ordering quantity of 1.5 times the previous amount, the ordering quantity of 1.75 times the previous quantity, and the ordering quantity of two times the previous quantity. Table 3 shows the simulation test results in several scenarios.

Table 3. Scenario simulation result comparison

Scenario	Average Ordering Quantity (Ton/month)	Average Inventory (Ton/Month)	Average Stock Shortage (Ton/Month)
I	3521	6217	151
II	3916	9628	75
III	4013	10943	0

The determination of the best result requires the understanding of planning and ordering the raw materials for the result without a stock shortage. In the scenario I, the ordering quantity of 1.5 times more than the previous amount had the average ordering quantity of 3521 tons/month and average inventory of 6217 tons/month while it experienced the stock shortage with an average of 151 tons/month.

Then for scenario II, the ordering quantity of 1.75 times more than the previous amount had the average ordering quantity of 3916 tons/month and average inventory of 6217 tons/month while the stock shortage was 75 tons/month in average. However, in scenario III, the ordering quantity of two times more than the previous amount had the average ordering quantity of 4013 tons/month and an average inventory of 10943 tons/month without the stock shortage.

Therefore, based on these three scenarios, the best scenario was scenario III with the ordering quantity of 4013 tons/month and the average inventory of 10943 tons/month. If the company applies this policy of the iron ore ordering and inventory, the company will minimize the possibility of a stock shortage each month

4. Conclusion

In conclusion, based on the system dynamics modelling and simulation, the result shows that the best solution was available for raw materials ordering and inventory. The CLD model represented the system behaviour of ordering raw material with several variables which related to each other, and four variable relations in the loop of raw materials ordering and production. The variables include Loop R1 which is production enhancement, Loop B1 which is iron ore ordering, Loop B2 which is iron ore inventory, and Loop B3 which is production.

Moreover, the SFD model, which represents the system behaviour showed the result of Oscillation. Oscillation is the condition of the production quantity which is often not adequate to satisfy the customer demand, in this case, the iron sand demand for each month. As the simulation result, the production quantity was within the range of 30 tons at a minimum and 200 tons as the maximum, and the iron ore utilization was within the range of 30 tons minimum and 240 tons as maximum.

The best scenario for raw materials ordering and inventory was to order two times more than the previous amount with an average quantity of 4.013 tons/month and an average iron ore inventory of 10.943 tons/month. If the company applies this scenario, it will minimize the possibility of the stock shortage each month. However, the company needs to consider additional suppliers to use this scenario.

5. Reference

[1] Andriansyah Sentia P D Elia S E Prasanti N 2018 Decision Making of Inventory System using Monte Carlo Simulation: a study case *International Journal of Conceptions on Computing and Information Technology* **6** 1 5-7

[2] Sterman J D 2000 *Systems Thinking and Modelling for a Complex World* (USA: The McGraw-Hill Companies, Inc)

[3] Botha A Grobler J Yadavalli V S S 2017 System Dynamics Comparison of Three Inventory Management Models in an Automotive Parts Supply *Chain Journal of Transport and Supply Chain Management* **11** 1-12

[4] Belhajali I and Hachicha W 2013 System Dynamics Simulation to Determine Safety Stock for a Single-Stage Inventory System *2013 International Conference on Advanced Logistics and Transport* 488-493

[5] Al-Refaie A Al-Tahat M and Jalham I 2010 A System Dynamics Approach to Reduce Total Inventory Cost in an Airline Fuelling System *Proceedings of the World Congress on Engineering* **1**.

[6] Verwater-Lukszo Z and Christina T S 2015 System-dynamics Modelling to Improve Complex Inventory Management in A Batch-wise Plant *Computer Aided Chemical Engineering*, **20** 1357-1362

[7] Banks J Carson J S and Nelson B L 1996 Discrete-Event System Simulation, 2nd Ed. (Englewood Cliffs: Prentice Hall)

[8] Robinson S 2003 *Simulation: The Practice of Model Development and Use* (Chichester: Wiley)

[9] Maria A 1997 Introduction to Modelling and Simulation *Winter Simulation Conference* 7–13.

[10] Qudrat-Ullah H 2012 On the Validation of System Dynamics Type Simulation Models *Telecommunication System* **51** 159–166

[11] Cedillo-Campos M and Sánchez-Ramírez C 2013 Dynamic Self-assessment of Supply Chains Performance: An Emerging Market Approach *J. Appl. Res. Technol* **11** 3 338–347

[12] Japar F Napitupulu H L and Siregar I 2013 Aplikasi Teknik Simulasi untuk Perencanaan Persediaan dan Pemesanan Bahan Baku di PT XYZ [The Implementation of Simulation Technique for Inventory Planning and Raw Material Ordering in PT XYZ] *Jurnal Teknik Industri USU* **3** 4 18–22.

IOP Conf. Series: Materials Science and Engineering **931** (2020) 012018 doi:10.1088/1757-899X/931/1/012018

Performance analysis using Fuzzy Balanced Scorecard-based performance prism

M R A Purnomo, K T Prayati
Department of Industrial Engineering, Faculty of Engineering Technology. Universitas Islam Indonesia, Yogyakarta, Indonesia

E-mail: ridwan_ie@uii.ac.id

Abstract. Managing an organisation without strategy is like as being blind in the world of globalisation. An organisation must formulate a new strategy and find out what parts or departments that should be prioritised to be improved more than others. To be able to formulate the new strategy that is adequate to improve employee performance, the company must be able to identify the source of the main problems that could be obstacles in improving employee performance. This paper presents development of hybrid method between Fuzzy Balanced Scorecard (FBSC) and Performance Prism (PP) to analyse performance of a company. This study took place at a multimedia company with big projects mostly from Indonesia and United States. Result of the study showed that the proposed hybrid method could be used to evaluate the company from 4 perspectives that are customer's satisfaction, finance, internal business process and learning and growth in PP framework. The fuzzy logic was incorporated in an analytical hierarchy process (AHP) to determine weight for every perspective based on experts' judgement. Result of this study showed that the investigated company has performance score very high that is 9.938, with the lowest score is employee stakeholder factor. With this result, the company would be able to track down the root cause of the problem and an effective strategic plan could be formulated in order to improve employee and overall company performance.

1. Introduction

Managing an organisation without strategy is the same as being blind in the world of globalization. Managers have to focus on their strategy to survive in competitive market [1]. In order to achieve their target, organisations have to improve their strategic plan because the economic condition sometime forces company to save the cost and made the allocation of funds more limited [2]. Therefore, an organisation must formulate a new strategy and find out which parts or departments that should be prioritised to be improved more than others, then allocate their limited resources such as budget, human resources, time and others to achieve the vision [3].

A company must be able to identify how the influence of strategy success or failure directly impacting on the objectives of the company [4]. The company's strategy is expected to focus on the level of performance or improvement as needed, as well as initiatives to get results of the needed program in order to achieve the target [5].

To be able to formulate the new strategy that is adequate to improve employee performance, the company must be able to identify the source of the main problems that could be obstacles in improving employee performance [5]. If a company able to identify such main problems, then the strategy to improve employee performance could be optimally formulated.

This study took place at a multimedia company with big projects mostly from Indonesia and United States. The main problem that occurs in the company is the lack of satisfaction that felt by most of the employees. The problem comes from the company services and benefit that they gave to the employees felt not fair and uncertain career development path that is supposed to be provided by the company. Besides that, the employee attendance and ambiguous attendance evaluation is also the problem to solve in order to improve the employee and company performance.

One of the methods that could be used to analyse the company and employee performance is performance prism (PP) that is coined by Neely et al [6]. However, improvement for analysis of every side in the PP is still required in order to get comprehensive analysis. In this study, comprehensive analysis owned by Balance Score Card (BSC) is incorporated in the PP so that the stakeholder satisfactions in PP would be analysed based on perspectives in the BSC. Next step are identification of Key Performance Indicators (KPIs) and weighting the identified KPIs. Since the KPIs weighting would be conducted based on experts opinion, then Analytical Hierarchy Process (AHP) was used as the method for KPIs weighting. It is commonly known that experts prefer to use linguistic variables in analysis. Those variables are also could reduce subjective opinion of the experts. However, for engineering analysis, numerical variables are required instead of linguistic variables. Therefore, fuzzy logic could be implemented in the proposed AHP to get quantitative weight of every KPI [7].

There are plenty of researches about BSC that have been done by previous researchers. Agrawal et al [1] used the BSC as the basis for determining the attributes and sub-attributes of the outsourcing organisation's performance appraisal to ensure that the organisation still can make maximum contribution with the use of reverse logistics. Other study was carried out by Lu et al, (2018) that integrated sustainability and BSC became sustainable BSC (SBSC) to evaluate performance of airport development [8].

In PM side, Nogning & Gardoni [9] has implemented PP using company's social responsibility theory and reflected that the first that must be considered is stakeholders. Previously, Youngbantao & Rompho [10] have conducted a research on the application of PP to measure performance in two companies with different types of culture; those were companies that have a culture of flexibility and a company with a culture of stability. The results showed the same level of performance effect in the two companies with different cultures [10].

Based on explanation above, it could be said that PP has successfully implemented to analyse performance of companies in any condition by focusing on stakeholders and still could be integrated with other methods to get better analysis tool. In another side, BSC has been accepted by performance analysis researchers as a comprehensive performance analysis tool. This study tried to incorporate BSC into PP in order to get advantages of BSC to be implemented in PP. Fuzzy-based AHP was also implemented in order to determine weight of every KPIs that have been evaluated by experts using linguistic variables

2. Theory

2.1. Performance Prism

Performance prism is a method of thinking that integrates five perspectives related to each other and provides a structure that allows executives to determine answers to five fundamental questions, those are stakeholder satisfaction, stakeholder contribution, strategy, process and capability [11]. Those five views altogether provide a comprehensive and integrated framework for managing organizational performance and solving organisational-related problems that could build a structured business performance model [12]. PP is currently considered more popular than the BSC, because it is considered to be a more flexible and more focused on stakeholders' method [13]. BSC's focus on stakeholders is limited to shareholders and customers only. Other stakeholder, such as employees, is not considered in BSC, hence, that is the advantage for PP because PP covers all stakeholders [14].

2.2. Balance Scorecard (BSC)

BSC method coined by Kaplan and Norton as a performance measurement tool for managers to find a quick and comprehensive picture of the company's operational performance [3]. BSC method combines several measurements, such as non-financial performance and traditional financial metrics to get balanced results from organisational performance. This assessment is used extensively to align business activities with the organisation's vision and strategy and to improve internal and external communication as well as monitoring business performance [15]. The success of the BSC depends on clear identification of non-financial and financial variables, accurate and objective measurement of the two variables and the relationship between performance improvement, rewards for performance achievement and penalties for performance violations [8].

BSC is more than just a business model because it able to facilitate performance measurement [3]. The broad multi-dimensionality of models and functions of the BSC system encourages different researchers and practitioners to study and analyse it from different perspectives. When performance evaluation is part of a safe management and control system, the organisation could effectively manage its resources and measure its performance in relation to the goals of the organisation [16].

3. Method and result

3.1. Hybrid framework design

Incorporating BSC into PP was carried out by inserted BSC perspectives into PP framework. One of the stakeholders in PP is customer, therefore, customer perspective was incorporated in customer stakeholder. Learning and growth perspective in BSC tries to evaluate ability of a company in learning to grow when answering external industrial challenges. Therefore, it affects to capability of the company. Because of that, learning and growth perspective in BSC was incorporated in capabilities factor in PP. Next perspective in BSC is internal business process, which tries to evaluate quality of business process in a company. This perspective was incorporated in process factor in PP. Actually, finance perspective is not the one and only one factor that could be used to measure performance of a company. However, in traditional performance measurement, finance still being used to measure performance of a company because it is believed that finance could be one of important parameter to justify the health of a company. In the proposed hybrid framework, finance was used as the perspective in the company performance analysis using PP after considering the customer satisfaction, internal business process and learning and growth perspective. Figure 1(a) shows the basic of PP model [17], while Figure 1(b) shows the modified PP model.

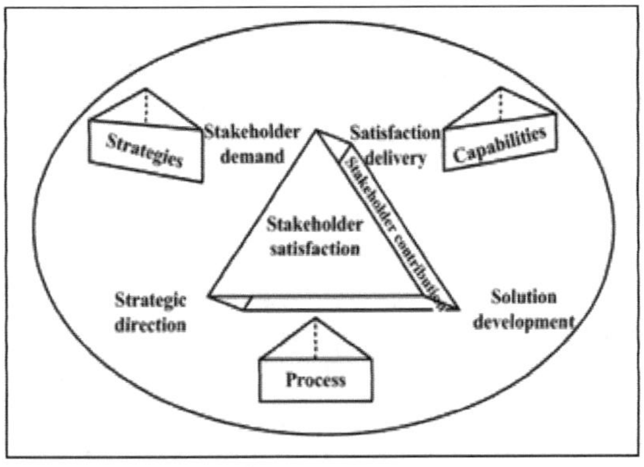

Figure 1a. Basic of PP model

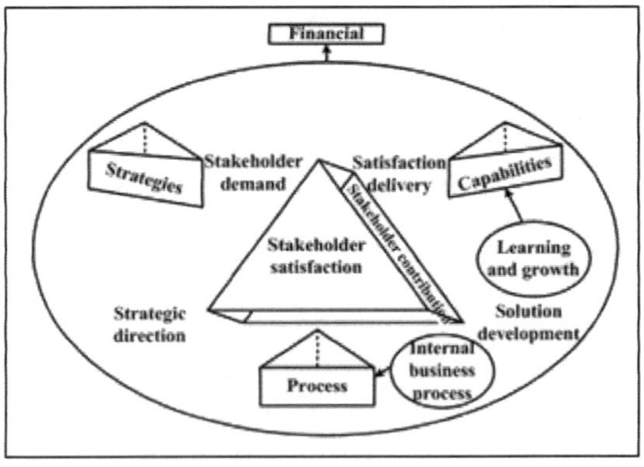

Figure 1b. Modified PP model

3.2. Data collection

The data is collected from year 2016 to year 2018. The first collected is financial data that used to identify pattern of benefit obtained by the company. The next collected data is number of projects that have been handled by the company. Customer satisfaction data was also collected through questionnaires that have been spread out to customers. Data about employee that are their resignation, attendance, satisfaction and productivity were also collected through spreading questionnaires, making in-depth interview and administration data stored in Human Resource department. The collected data were then been processed and mapped into the related perspectives in the modified PP.

3.3. Key Performance Indicators (KPI) identification and validation

The KPI identification and validation consist of few processes, those are:

1. Classifying stakeholders of the investigated company, which are CEO, executive management, client, customer, and employee.
2. Identifying the satisfaction, contribution, strategy, process and capability of each stakeholder.
3. Design KPIs that are adjusted to the business goals of each stakeholder.
4. The KPIs were validated through discussions with competent management. Thus, the KPIs that have been designed could describe actual overall performance of every aspect of the company

3.4. Weighting and scoring KPI with Fuzzy-AHP method

After validating the KPIs, they would be weighted using Fuzzy-AHP method. This process was performed with online software for fuzzy-AHP, which is Online-Output software. After weight of each KPI was obtained, then each KPI would be scored based on company achievement in an interval of time. Determination of the highest and the lowest value of each KPI were carried out through interview and discussions with relevant management. Table 1 shows result of performance score of each stakeholder using fuzzy-AHP.

Table 1. Performance score of each stakeholder

Stakeholders	Weight	Priority	Performance Value	Average Performance Value	Overall Performance Value
CEO	0.573	1	50.845	10.169	5.827
Executive Management (EM)	0.084	4	51.320	10.264	0.862
Client	0.200	2	45.000	9.000	1.800
Customer	0.118	3	52.165	10.433	1.231
Employee	0.025	5	43.580	8.716	0.218
				TOTAL	9.938

3.5. Company's total performance analysis

In order to evaluate total performance of the company, then an evaluation standard has been arranged based on discussion with managers in the company. The performance standard is as follows:

- Performance value 1-5 : underperformed
- Performance value 6-10: adequate
- Performance value 11-15: advance

Based on the overall performance value, which is 9.938, it could be said fairly that performance of the company is adequate and close to upper limit. However, goal of the company is to achieve advance performance, it means that there are several KPIs that should get special attention and be improved according to the company's target. Besides, from the performance score, it could be seen that the employee stakeholder has the lowest weight and score of overall stakeholder performance. It means that the company should consider the employees to be more important in the role of stakeholders and give more attention to the employees and provide them with various ways to improve their performance.

3.6. Root cause analysis for underperformed KPI

Figure 2 shows the map of root cause analysis for underperformed KPI.

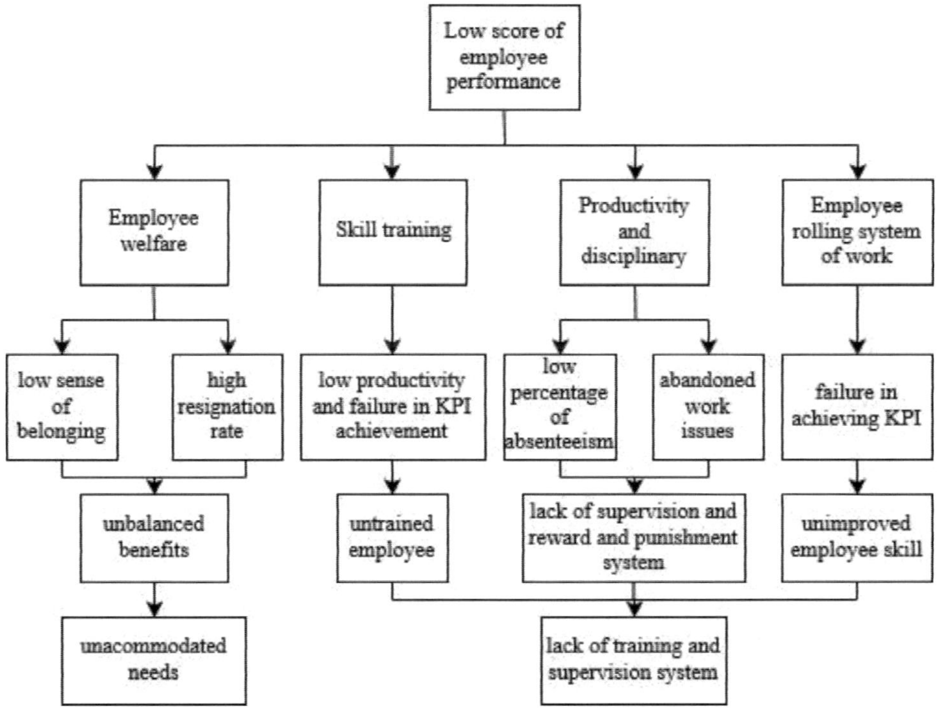

Figure 2. Map of root cause analysis for underperformed KPI

From the Figure 2, it could be seen that main problems that makes the employee performance score is low are un-accommodated needs of employee and lack of training and supervision system from the company to employees. Several suggestions for improvement are as follows:

1. Allocate some capital to have effective employee welfare program, where the employee could share their needs and it will be easier for the company to accommodate the employee needs.
2. Organise trainings and also evaluate how effective the training in terms of improving employee performances. The training could be held periodically, it could be once every semester or once every three months, according to the company's needs and budget.
3. Implement an effective reward and punishment system in order to improve discipline of the employee. It is suggested to the company to have evaluation form for the discipline that would be filled in by HR department based on the attendance system. The form could be evaluated monthly to decide reward or punishment that should be given to an employee.
4. The team leaders should implement a task-rolling system for their team, in which each of the team members will try to excel one task for few months and then each of them will train the other team members, and let them try to excel another task. While doing that, the leaders should have their own evaluation method in which they will evaluate each performance of the team members, they will have the back up for each task and they shouldn't be worry anymore of the failure in achieving KPIs.

4. Conclusion

From the discussion above, it could be concluded that the proposed hybrid system, which is incorporated BSC into PP has successfully been implemented to analyse performance of a service industry. Fuzzy-AHP was also successfully used to do scoring for KPIs of every perspective in BSC. The use of fuzzy could reduce subjective opinion from experts because linguistic variables were used for evaluation instead of numerical variables. Result of the proposed hybrid system could be continued with root cause analysis in order to identify root problem of underperformed KPIs. Hence, effective actions could be suggested to overcome the root problem and further could improve the overall company performance.

5. Reference

[1] Agrawal S Singh K R and Murtaza Q 2016 Outsourcing Decisions in Reverse Logistics: Sustainable Balanced Scorecard and Graph Theoretic Approach *Resources, Conservation and Recycling* **108** 1 41-53
[2] Kalender T Z and Vayvay O 2016 The Fifth Pillar of the Balanced Scorecard: Sustainability *Procedia - Social and Behavioral Sciences* **235** 1 76-83
[3] Lin H Y Chen C C Tsai M F C and Tseng L M 2014 Balanced Scorecard Performance Evaluation in A Closed-loop Hierarchical Model Under Uncertainty *Applied Soft Computing* **24** 1 1022-1032
[4] Babanova Y Dolinskaia A and Kolyada A 2018 *The Main Features of Strategic Business Consulting in the Conditions of Digital Economy*
[5] Senvar M O Vayvay O Kurt E and Hloch S 2014 Prioritization of Balanced Scorecard Measurement Indicators as A Process Management Approach via Fuzzy AHP: Case Study in Automotive Industry *Tehnički vjesnik* **21** 1 155-162
[6] Neely A Adams C and Kennerley M 2002 The Performance Prism *The Scorecard for Measuring and Managing Business Success* (Pearson Education Limited)
[7] Perez A C Montequin R V Fernandez O F Balsera V J 2017 Integration of Balanced Scorecard (BSC), Strategy Map, and Fuzzy Analytic Hierarchy Process (FAHP) for a Sustainability Business Framework: A Case Study of a Spanish Software Factory in the Financial Sector *Sustainability Journal* **527** 9 1-23
[8] Lu T M Hsu C C Liou J H L and Lo W H 2018 A Hybrid MCDM and Sustainability-Balanced Scorecard Model to Establish Sustainable Performance Evaluation for International Airports *Journal of Air Transport Management* **71** 1 9-19
[9] Nogning F R and Gardoni M 2017 Double Performance Prism: Innovation Performance Measurement Systems for Manufacturing SMEs *International Journal of Innovative Research in Advanced Engineering (IJIRAE)* **4** 03 113-128
[10] Youngbantao U and Rompho N 2015 The Uses of Measures in Performance Prism in Different Organizational Cultures *Journal of Accounting and Finance* **15** 6 122-130
[11] Liu J Love P E D Smith J Regan M and Palaneeswaran E 2014 Review of Performance Measurement: Implications for Public–Private Partnerships *Built Environment Project and Asset Management* **5** 1 35-51
[12] Estrada R Sousa S Lopes I 2017 Performance Assessment System Development Based on Performance Prism in Social Services *Proceedings of the 2017 IEEE IEEM* **17** 1 854-859
[13] Smulowitz S 2015 Evidence for the Performance Prism in Higher Education *Measuring Business Excellence* **19** 1 1-21
[14] Regha R S and Rani U R 2018 A Fuzzy based Enhancement on Prism and J48 Classifier Prediction of Student Performance *International Journal of Advanced Technology and Engineering Exploration* **5** 42 2394-7454
[15] Severgnini E Galdamez E V C and Moraes R O 2017 Satisfaction And Contribution Of Stakeholders from the Performance Prism Model *Brazilian Business Review* 121-136
[16] Zhao H Li N 2015 Evaluating the Performance of Thermal Power Enterprises using Sustainability Balanced Scorecard, Fuzzy Delphic and Hybrid Multi-criteria Decision-Making Approaches for Sustainability. *Journal of Cleaner Production*
[17] Shen C Y Chen S P Wang H C 2015 A Study Of Enterprise Resource Planning (ERP) System Performance Measurement using the Quantitative Balanced Scorecard Approach *Computers in Industry*

IOP Conf. Series: Materials Science and Engineering **931** (2020) 012019 doi:10.1088/1757-899X/931/1/012019

Development of rule-based expert system for conceptualisation of poster design

M R A Purnomo, S Hidayatuloh
Department of Industrial Engineering, Faculty of Industrial Technology, Universitas Islam Indonesia, Yogyakarta, Indonesia

E-mail: ridwan_ie@uii.ac.id

Abstract. This study discusses development of rule-based expert system (R-ES) as a decision support for poster design conceptualisation. Users of the proposed expert system are novice poster designers while the experts are senior and professional poster designers. Development of R-ES is started by defining type of poster to be designed. Based on professional poster designer's experiences, 5 types of poster have been proposed, which are advertising, information, movie, event and political poster. A knowledge-based system consists of IF-THEN rules have been developed by the experts to determine poster design concept that are colour combination, poster size, font type and size, expert note, and poster sample. Result of this study is an online expert system for conceptualisation of poster design that could be widely used by novice poster designers without constrained by space and time.

1. Introduction

According to Bureau of Labour Statistic, demand of graphic designer will grow up to 4% in 10 years ahead. Other than that, there is growing about 24-26% in computer field or digital area [1]. Graphic design is a practical activity consists of visual material, typography, pictures and other elements to communicate or create experiences to the audiences [2]. Graphic designer needs to understand how to communicate a message through graphic design. There are some common issues that novice graphic designers usually do, such as "design for themselves", "overestimating the client's ability to translate words into visual language", "making a similar design as existed design" and "less knowledge in technical design". On the contrary, professional poster designer would digs more information about client needs, conduct deep researches in design scope and takes other designer's experiences for consideration [3]. Thus, novice designers need professional designer's experiences in visualising the message through graphic design.

PW is a small to medium company that runs their business in digital printing and produces graphic design product such as poster, banner, pamphlet and others. Graphic design is the crucial process in this company; unfortunately they have faced issues. Some of products require more time to be finished, because the prototype does not meet the client's expectation, then a lot of design revisions should be carried out. Designing is not only about splendid design but also how to make it works [4]. All of the graphic designers who work in this company are senior high school graduated. They have not attended proper graphic design school before. They can operate the software to design by self-taught but do not know more about visual communication knowledge. The company owner cannot hire professional graphic designer because of the high salary demand.

Motivation of this study is to extract professional poster designers' knowledge and experiences to be a systematic knowledge that could be used as reference when making poster design. In term of the knowledge based system development, professional poster designers would be the experts who have special knowledge or expertise that most humans do not have [5]. In this study, based on the professional poster designers' opinion, there are 5 types of poster that would be covered. The proposed system would be an expert system (ES) and user of the ES is novice poster designers. Extracted knowledge from the professional poster designers would be represented using a rule based system that consists of IF-THEN rules; therefore the proposed ES is called as rule-based ES (R-ES). In term of decision making, an ES would make decision based on subjective expert opinion. Therefore, sometime ES provides less accurate solution. However, an ES is able to provide accurate solution when expert's knowledge could be acquired well. In previous studies, expert systems played a role in detecting optometrists in livestock, where the results of detection by expert systems has high accuracy as well as diagnosis that made by an expert [6, 7].

One of the advantages of ES compared to human experts is ES could work 24 hours per day and 7 days per week. Besides, ES also able to conduct computation in very short time. In the investigated company, demand of new poster design is very high. Poster designers in that company mostly are novice poster designer; therefore, ES for poster design conceptualisation would plays an important role.

2. Literature survey

2.1. Graphic design

Graphic design could be described as a work of art that has functions, aspects of beauty, and a regular structure for several different types of text and illustrations or images. The purpose of graphic design is to find the right presentation to convey the message according to the recipient, the media and the economic situation [8]. One of the important factors in a poster is the graphic design. There are several previous studies that showed the importance of graphic design in delivering messages to the audiences or users. Soylucicek [9] conducted a study on educational games with the application of graphic design to attract children's attention through visual appearance of the game. That study proofs the important role of graphic design in delivering information. Another study conducted by Shahdad & Filmore [2] in the field of graphic design using TRIZ method that has been combined with a well-known optimisation technique that is Genetic Algorithm (GA). Altis & Ekren [10] and Suharjito et al. [6] conducted similar study that used expert systems to solve problems, a particular problem and proved that the proposed ES was a good method for engineers to solve problems [11]. Based on reviews above, it is known that there have been several graphic design studies with different methods, yet there were no previous studies in the field of graphic design by applying an ES as decision support. Therefore, this study aims to develop decision support systems for poster design using R-ES.

2.2. Expert system (ES)

Expert System is a computer intelligence program that uses knowledge and procedures to draw conclusions from existing facts to solve difficult problems using significant expertise. Two parts of expert system are the development system and the consultation environment. The development system or development environment is the part where the expert system is built. The consultation environment describes how the instructions or recommendations are given to the user. The development process begins with knowledge engineering that builds the system by gaining knowledge from experts. This knowledge is programmed on the knowledge base as facts about the subject area (domain), usually under conditions of 'IF-THEN' rules [5]. Architecture of the proposed R-ES in this study is depicted in Figure 1.

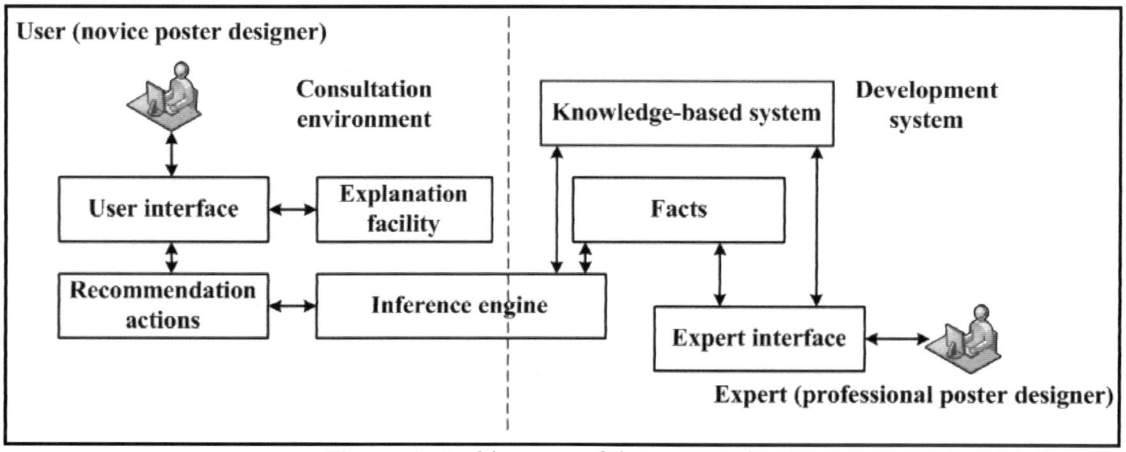

Figure 1. Architecture of the proposed R-ES

2.3. Role of graphic design in poster design

Posters are two-dimensional products, a one-page formatted used to inform or to influence or to promote something. The main purpose of making posters is to communicate or deliver messages [12]. Poster must have some basic compositions as follows: 'grab attention', 'set it apart', 'communicate key message', and 'single surface, one unit' [13]. The compositions would be delivered through a graphic; therefore, graphic design becomes critical thing in poster design.

3. Methods

3.1. Data collection

In this study, primary data is collected through in-depth interview with the professional graphic designers. Objective of the interview is to get basic knowledge on poster design, especially on graphic design. Even though the developed R-ES has interface for the professional graphic designers to add, edit or delete the facts and the knowledge-based system, however, the R-ES requires initial facts and knowledge that could be used to design the database for the R-ES. Secondary data is collected from internet. The secondary data that have been collected such as poster design, type of poster design and color imagination in graphic design.

3.2. Knowledge acquisition

The knowledge base that required for poster design is information about the types of posters, fields or types of industries that are intended for making posters, sub-fields or sub-categories that are intended for poster making, the atmosphere or mood displayed on posters and areas where posters will be shown or exhibited.

3.3. Knowledge-based system construction

There are three steps in constructing the rule that are symbolisation, decision tree and rule construction. Decision tree was created by analysing existing expert knowledge in the application domain. There several elements used to design a poster, which are colour, image, typography and poster size. From these elements, categorisation is developed to design decision trees. Categorisation is built from variables that arise from the problem domain. Before making decision tree, it is required to transform the variables into symbols. In this study, variables for poster design are question, poster type, category, sub variable, mood, area, and gender object variable. Table 1 shows the X symbols and their meaning. Table 2 shows F symbols that represent display area of poster and Table 3 shows the Y symbols that represent poster design output. Figure 2 shows simplified decision tree of the proposed R-ES.

Table 1. X symbols and their meaning

No	Code	Question Variables
1	X1	What type of poster do you want to design?
2	X2	What kind of industry or category do you want to design the poster?
3	X3	What kind of sub category do you want to design the poster?
4	X4	What kind of mood that you desire to express in your poster?
5	X5	Who is you want to show the poster?
6	X6	Where do you want to show your poster?

Table 2. F symbols and their meaning

No	Code	Area
1	F1	Indoor
2	F2	Outdoor
3	F3	Social media

Table 3. Y symbols and their meaning

No	Reasoning Rule
1	*If* Advertising Poster *and* Child product *and* love *and* playful *and* Indoor *then* Y1
2	*If* Advertising Poster *and* Child product *and* love and playful *and* Outdoor *then* Y2
3	*If* Advertising Poster *and* Child product *and* love and playful *and* Social Media *then* Y3
4	*If* Advertising Poster *and* Child product *and* joy and calm *and* Indoor *then* Y4
5	*If* Advertising Poster *and* Child product *and* joy and calm *and* Outdoor *then* Y5
6	*If* Advertising Poster *and* Child product *and* joy and calm *and* Social Media *then* Y6
7	*If* Advertising Poster *and* Child product *and* happiness *and* Indoor *then* Y7
8	*If* Advertising Poster *and* Child product *and* happiness *and* Outdoor *then* Y8
...	...
336	*If* Political Poster *and* trust *and* Social Media *then* Y336

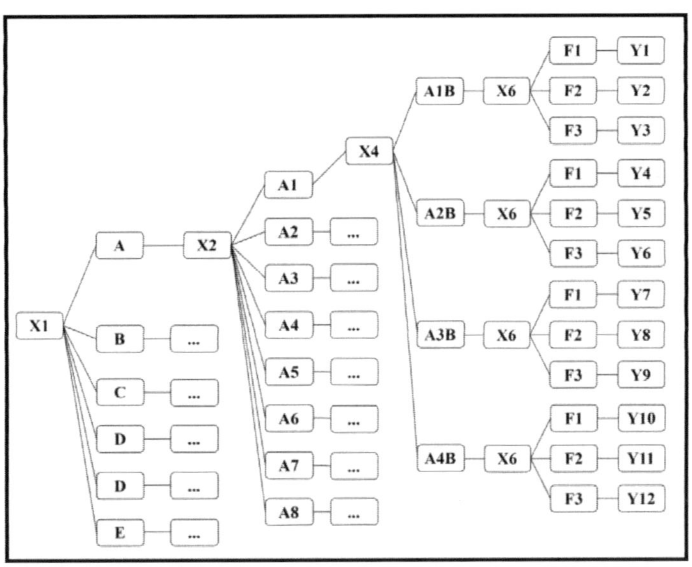

Figure 2. Simplified decision tree of the proposed R-ES

3.4. Entity Relationship Diagram (ERD)
Relationship between objects in the proposed R-ES is shown in Figure 3.

4. Result and discussion
Result of this study is a R-ES. First, users or experts must login to the system so that the R-ES could prepare related modules and menus. Figure 4 shows the login page for users and experts. For user's side, after logged in, the user would be able to add parameters and answer the added parameters in order to get concept of the poster to be designed. Figure 5 shows the menu to add the parameters while Figure 6 shows interface to answer every parameters by the user. After user finished answering all of the parameters then the R-ES would generate the concept of the poster to be designed and a design example. Figure 7 shows the interface to display the concept of poster design that covers colours, font type, poster size, general note, and poster sample. By reading the generated concept and seeing the poster example, the user would be able to generate ideas to design a new poster. Figure 8 shows example of poster design by novice poster designer after being assisted by the proposed R-ES. Type of the poster is indoor poster for advertising of children's product with joy theme with soft colour.

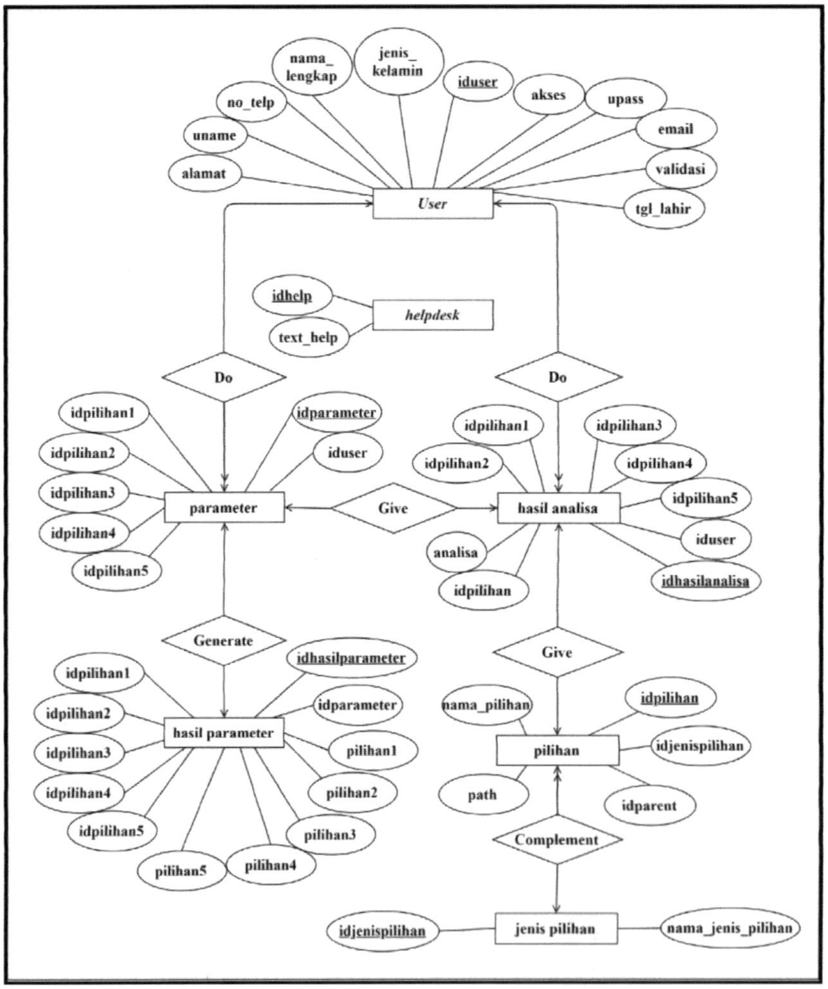

Figure 3. ERD of the proposed R-ES

Figure 4. Login page for users and experts

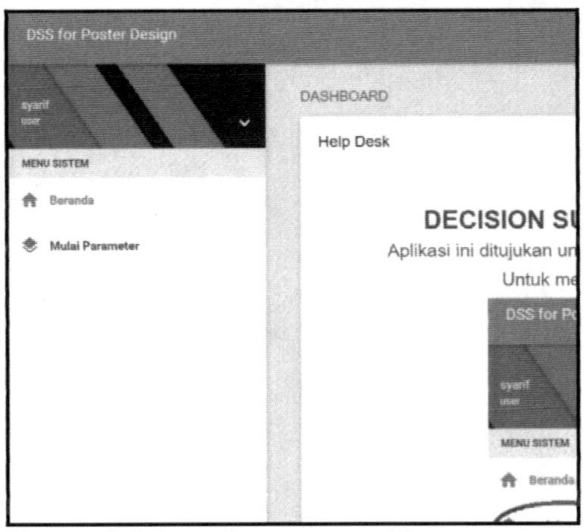

Figure 5. Menu for user to add parameters of poster design

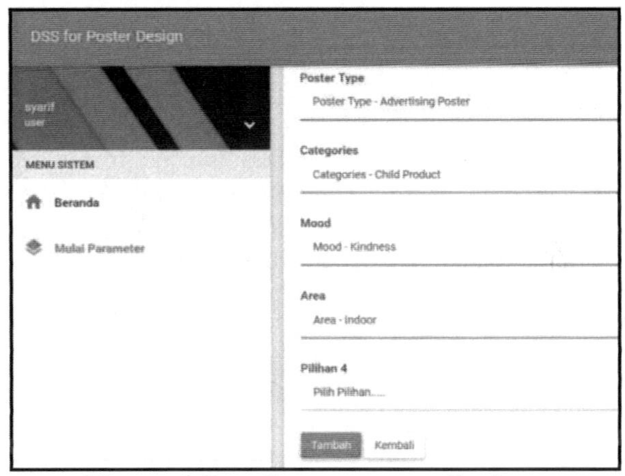

Figure 6. Interface to answer all of parameters of poster design

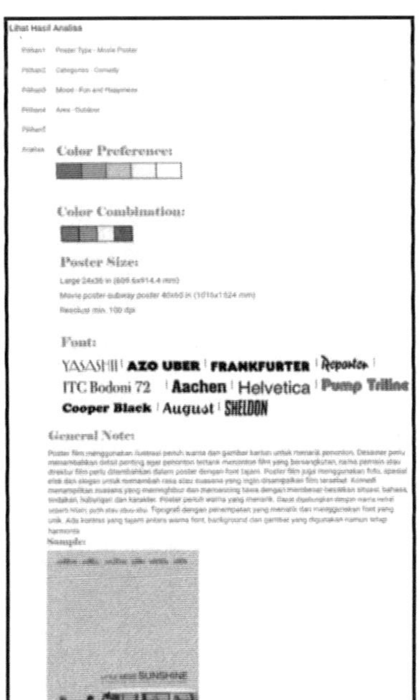

Figure 7. Interface to display concept for poster design

Figure 8. Example of poster design by novice poster designer assisted by proposed R-ES

5. Conclusion

Based on the explanation above, it could be concluded that the developed knowledge-based system that consists of IF-THEN rules could be used to represent professional poster designer knowledge. The developed knowledge-based system could be extended to be a R-ES and could be used to assist novice poster designer in designing a poster. The generated concept of poster design is also not very rigid. It means that the generated concept of poster design is just used to trigger creativity of the poster designer. Therefore, another benefit of the proposed R-ES is it could be used to train novice poster designers to be professional poster designers.

6. References

[1] Bureau of Labor Statistics 2018 *Occupational Outlook Handbook* (Washington D C: US Department of Labor)

[2] Shahdad M A and Filmore P 2015 Applying TRIZ to Graphic Design using Genetic Algorithms *Procedia Engineering* **131** 881 – 891

[3] Laing S and Apperley M 2017 Investigating the Effects of Client Imagery on the Ideation Process of Graphic Design 0142-694X *Design Studies* **53** 78-98.

[4] John J 2012 *Problems Faced by Graphic Designers* [Online] Available at https://suyati.com/blog/problems-faced-by-graphic-designers/ Accessed on August, 23, 2018

[5] Turban E Rainer R K Potter R E 2001 *Introduction to Information Technology* (United States of America: John Wiley & Sons, Inc)

[6] Suharjito Diana Yulyanto Nugroho A 2017 Mobile Expert System Using Fuzzy Tsukamoto for Diagnosing Cattle Disease *Procedia - Computer Science* **116** 27-36

[7] Ibrahim F Mohamad M I Makhtar S N Ibrahim J 2007 Classification of Risk in Dengue Fever and Dengue Haemorrhagic Fever using Rule Based Expert System *IFMBE Proceedings* **15** 50-53

[8] Pettersson R 2015 *Information Design 4: Graphic Design* (Austria: International Institute for Information Design)

[9] Soylucicek S 2012 Graphic Design on Educational Computer Games *Procedia - Social and Behavioral Sciences* **46** 2083 – 2087

[10] Altis S and Ekren N 2016 Development of an Outdoor Lighting Control System Using Expert System *Energy and Buildings* **130** 773–786

[11] Yang Y Fu C Chen Y W Xu D L and Yang S L 2016 A Belief Rule Based Expert System for Predicting Consumer Preference in New Product Development *Knowledge-Based Systems* **94** 105-113

[12] Yan L and Wang X 2018 Why Posters Contribute Different Content in Their Positive Online Reviews: A Social Information-Processing Perspective *Computer in Human Behavior* **82** 199-216.

[13] Landa R 2014 *Graphic Design Solution* 5th Edition (USA: Wadsworth)

Taguchi method application to improve the quality of coffee pulp screen printing products

S Zuhri[1], Ilyas[1], M Winanda[1], N Izzaty[1]
[1]Industrial Engineering Department, Universitas Syiah Kuala, Darussalam, Banda Aceh, Indonesia

E-mail: sarika.zuhri@unsyiah.ac.id

Abstract. One of the main reasons consumers choose a product is based on quality. Consumers will choose products that have the quality as expected and in accordance with the benefits obtained. This is the basis for conducting research on Coffee Pulp Screen Printing products. The study was conducted with the Taguchi Method to determine the optimal combination of color parameters produced by coffee pulp screen inks. The parameters studied were Coffee Pulp as a factor A, the amount of Water as a factor B, the amount of vinegar as factor C and the length of cooking time as a factor D. experiments were carried out with two levels and two values for each factor. In this experiment using orthogonal array L8 (27) with two replications. The results of the analysis were processed using S/N average and variability ratios. Based on the result of these experiments it was found that the factors that significantly influence the color quality of coffee plup screen printing ink are coffee plup 0,5 kg, o,5 liter water, 100 ml vinegar, and 15 minutes cooking time. The results of the experiment show the optimal combination to get a dense color on the results of coffee grounds.

1. Introduction

Garment products are one of the commodities with great potential to be developed in the global market. Since 2010, exports of apparel from Indonesia abroad, especially to Europe, have increased again [1]. That process is an opportunity for the garment industry in Indonesia to develop into a flagship. One of the growing garment businesses is the screen printing business. This business has the potential to become a superior business, because it is supported by the needs and awareness of the community's appearance.

One of the natural ingredients used for making screen printing inks is Coffee Pulp. Coffee pulp is used as the basis for making screen printing ink because it has a deep black color and is suitable as a base material for ink. Coffee pulp is waste disposal from the results of making or filtering coffee. Coffee screen printing ink has a composition ranging from coffee grounds, vinegar and water. Ink pulp coffee grounds itself is a water-based screen printing ink and consists of natural ingredients so there are no harmful ingredients contained therein.

2. Methods

Coffee pulp screen printing products are products that are still in research and development, of course there are still some problems that arise in the results of coffee pulp screen printing ink. One of the problems is the lack of color produced by the screen printing ink on the screen printing results due to the lack of proper standardization of the composition for coffee pulp screen printing inks. problem

IOP Conf. Series: Materials Science and Engineering **931** (2020) 012023 doi:10.1088/1757-899X/931/1/012023

solving for this coffee pulp screen printing product can be done with the help of the taguchi method which serves to determine the most influential factors and the best combination to improve product quality.

The taguchi method itself is a method that has the aim of improving product quality at the same time reducing costs and resources to a minimum. The taguchi method is also an efficient method because it is possible to carry out research involving many factors and quantities and produce conclusions regarding the response of the factors and levels of control factors that produce the optimum response [2].

Based on previous research, taguchi method is a method that can improve quality, productivity [2] and also the method has aims to reduce costs and resources to a minimum [3]. Taguchi method is also very effective in determining the right combination of factors or composition of a product, because it involves many factors [4][5][6].

The steps in implementing the taguchi method according Soejanto [2] are.

1. Formulate / define the problem that will be investigated in the experiment.
2. The goals that underlie the experiment must be able to answer what has been stated in the formulation of the problem, namely looking for the cause that becomes the effect on the problem that we observe. This search is done systematically
3. Identification of Factors (Independent Variables) using causal diagrams.
4. Separation of Control Factors and Disruption Factors
5. Determination of the Number of Levels and Level Factor Values, this is important for
6. The accuracy of the results of the experiments and the costs of conducting the experiments.
7. Calculation of Degrees of Freedom
   Dof for factor A = nA - 1
   Dof for the interaction of factors A and B = (nA - 1). (NB - 1)
   Total number Dof = (nA - 1). (NB - 1) + (nA - 1). (NB - 1) (1)
8. The choice of an appropriate orthogonal matrix depends on the factor and interaction values expected and level values of each factor.
9. Placement of Columns for Factors and Interactions into the Orthogonal Matrix
10. Preparation and Implementation of Experiments. Implementation of experiments includes determining the number of replication experiments and randomization of the implementation of experiments.
    a. Number of Replication carried out with the aim of:
    - Increase accuracy of experimental data
    - Reduces the error rate in experiments
    - Obtain an estimated price of experimental error so as to enable a significant test of experimental results to be held.

    b. Randomization
11. Analysis of the Taguchi Variant. For a two-way analysis of variance the requirement that must be met is that the experimental data must consist of two factors or more and two or more levels.
    S_T - Total Squares.
    $$SS_T = \sum_{i=1}^{n} y^2 \quad (2)$$
    S_A - Number of Factor A Squares
    $$SS_A = \left[\sum_{i=1}^{KA} \frac{A_i^2}{n_{A_i}} - \frac{T^2}{n}\right] \quad (3)$$

S_{AxB} - number of AxB interactions

$$S_{AxB} = \frac{(TotalAxB1)^2}{n_1} + \frac{(TotalAxB2)^2}{n_2} - \frac{(TotalAxB)^2}{n_1 + n_2} \quad (4)$$

SSe - Number of Squares error.

$$SS_T = SS_A + SS_B + SS_{AxB} + SS_e$$

$$SS_e = SS_T - SS_A - SS_B - SS_{AxB} \quad (5)$$

The results of the analysis of variance do not prove the existence of differences in treatment and the influence of factors in the experiment, this proof is carried out hypothesis test F.

$$F_{source} = \frac{\text{variance due to treatment + error variance}}{\text{variance due to error}} \quad (6)$$

The pooling up strategy was designed by Taguchi to estimate the error variance in the analysis of variance. S / N ratio S / N ratio (Signal to Noise ratio) is used to select factors that have contributed to reducing the variation of a response. The S / N ratio used is that the smaller the quality characteristics, the better.

$$S/N = -10\log\left(\frac{1}{n}\sum_{i=1}^{r} Y_i^2\right) \quad (7)$$

3. Result and discussion

Before determining the orthogonal array matrix model that will be used, first determine the control factor by identifying the independent variables. Determination of control factors using a fishbone diagram, as shown below:

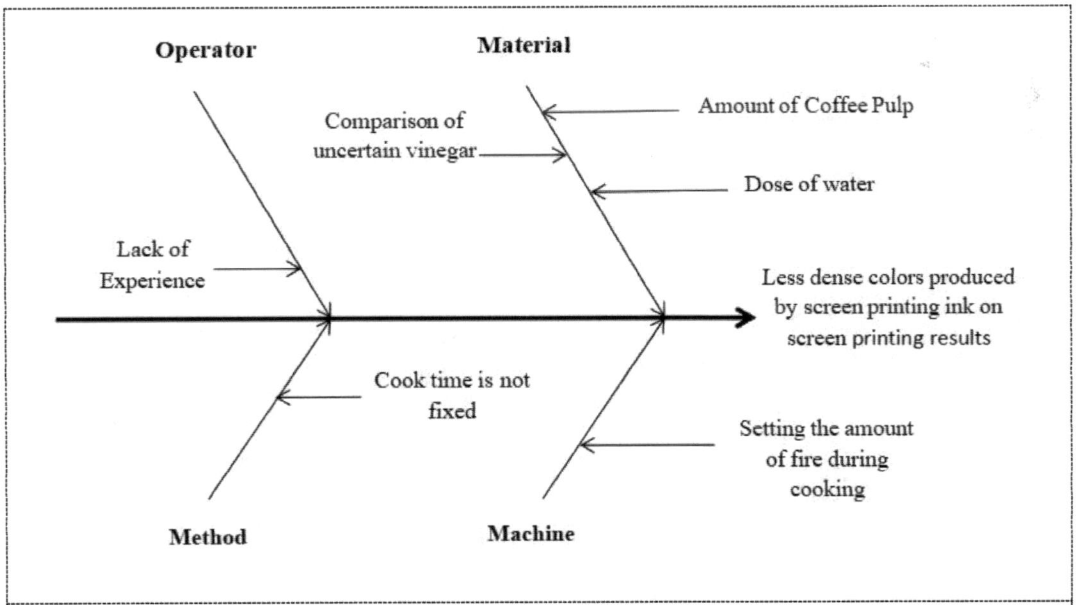

Figure 1. Fishbone Diagram.

The control factors that are determined based on the control factors that have a level value include: coffee pulp (A), water (B), vinegar (C) and cooking time (D).

Table 1. determining the number of levels and factor level values

Code	Control Factor	Level 1	Level 2
A	Amount of Coffee Pulp	0,5 Kg	1 Kg
B	Dose Of Water	0,5 Litre	1 Litre
C	Amount Of Vinegar	100 ml	200 ml
D	Cooking Time	10 Minutes	15 Minutes

Based on the control factors that have been obtained. there are several interactions between the factors that occur are the interaction of factors AxB, factors AxC and factors AxD.

The orthogonal array chosen is $L_8(2^7)$, with a matrix like in table 2. Table 3 shows in the recording there are two replications that are the same in an experiment with the same conditions to obtain more accuracy high on coffee pulp screen printing inks. The experimental results are obtained by conducting an experiment related to the four factors that were previously created according to a combination factor level matrix. This is done to obtain a more accurate estimate of the effect of a factor is done repetition (replication).

Table 4 shows the ranking of the factors of each level, after calculating the SNR difference of the two levels to determine the ranking of each factor. The factor that has the biggest difference is ranked first and so on. The ranking states how much influence or contribution to product quality Based on the calculation of Signal to Noise Ratio (SNR) can be obtained that the interaction between factors AxB level 1 has a large influence or contribution to the experimental results with an SNR value of 11,67 and less influential is the C level 1 factor with an SNR value of 11,13.

Table 2. Orthogonal Array Matrix

Matriks Orthogonal Array							
Eks	1	2	3	4	5	6	7
	A	B	AxB	C	AxC	AxD	D
1	1	1	1	1	1	1	1
2	1	1	1	2	2	2	2
3	1	2	2	1	1	2	2
4	1	2	2	2	2	1	1
5	2	1	2	1	2	1	2
6	2	1	2	2	1	2	1
7	2	2	1	1	2	2	1
8	2	2	1	2	1	1	2

Table 3. Results of Experiments on Coffee Silk Screening Ink Products

Matriks Orthogonal Array											
	Factor and Interaction							replication			
Eks	1	2	3	4	5	6	7	1	2	total	Mean
	A	B	AxB	C	AxC	AxD	D				
1	1	1	1	1	1	1	1	9	10	19	9.5

2	1	1	1	2	2	2	2	9	7	16	8
3	1	2	2	1	1	2	2	6	5	11	5.5
4	1	2	2	2	2	1	1	7	8	15	7.5
5	2	1	2	1	2	1	2	6	7	13	6.5
6	2	1	2	2	1	2	1	7	5	12	6
7	2	2	1	1	2	2	1	8	8	16	8
8	2	2	1	2	1	1	2	8	6	14	7
					Average						7.25

Table 4. Ranking factors based on S / N ratio

Factor/ Level	A	B	A x B	C	A x C	A x D	D
Level 1	11,30	11,10	11,67	11,13	11,29	11,50	11,40
Level 2	10,87	10,94	10,51	11,04	10,88	10,83	10,77
Difference	0,43	0,16	1,16	0,09	0,41	0,67	0,63
Ranking	4	6	1	7	5	2	3

Table 5. Analysis Of Variance

Sumber	DoF	SS	MS
A	1	1,23	1,23
B	1	0,58	0,58
AxB	2	8,19	4,09
C	1	0,05	0,05
AxC	2	1,17	0,58
AxD	2	0,55	0,27
D	1	2,69	2,69
Error	1	4,32	4,32
Total	11	18,78	-

Analysis of variance calculation of the number of squares error is used to find out the error that occurred obtained from the difference in the total number of squares with all average square and number of factor squares [7]. The results of these calculations obtained an error of 4,42 which is used to calculate the error in the analysis variance. With an error of 4,32 and a degree of freedom 1 an average square error is obtained (mean square) of 4,32.

Table 6. Polling Up Factors I

Code	V	SS	MS
A	1	1,23	1,23
B	1	0,58	0,58
AxB	2	8,19	4,09
C		Pooling factor C	
AxC	2	1,17	0,58
AxD	2	0,55	0,27
D	1	2,69	2,69
Error	2	4,37	2,18
Total	11	18,78	-

Table 7. Polling Up Factors II

Code	V	SS	MS
A	1	1,23	1,23
B	1	0,58	0,58
AxB	2	8,19	4,09
AxC	2	1,17	0,58
AxD		Pooling Interaction Factor AxD	
D	1	2,69	2,69
Error	4	4,92	1,23
Total	11	18,78	-

Table 8. Polling Up Factors III

Code	V	SS	MS
A	1	1,23	1,23
B		Pooling factor B	
AxB	2	8,19	4,09
AxC	2	1,17	0,58
D	1	2,69	2,69
Error	5	5,50	1,10
Total	11	18,78	-

Table 9. The results of the analysis of variance incorporating "polling up factors"

Code	V	SS	MS
A	1	1,23	1,23
AxB	2	8,19	4,09
AxC	Pooling Interaction Factor AxC		
D	1	2,69	2,69
Error	7	6,67	0,95
Total	11	18,78	-

Tables 6.7.8.9 show the pooling done to collect insignificant factors are referred to as errors. Factor said no significant if the lowest SS value is obtained, Table 6.7.8.9 shows that factors C, D, AxD and AxC are not significant and is considered an error because the SS values of factors C, D, AxD and AxC have the smallest values so that the factors C, D, AxD and AxC are pooled up. By pooling factors C, D, AxD and AxC then obtained MSeror of 0.95 which was previously 4.32, which means a decrease error value.

Analysis of calculation of percent contribution is used to find out how much percent contribution of each factor to the quality of coffee pulp screen printing products [8]. The calculation of percent contribution is obtained from the comparison of the SSfactor value against SSTotal. From the calculation of the percent contribution, it was found that the interaction between factor A (coffe pulp) and B (water) had the largest contribution, namely 33,49%.

Table 10. Percent Distribution

Code	DoF	SS	MS	SS'	P%
A	1	1,23	1,23	0,28	1,49
AxB	2	8,19	4,09	6,29	33,49
D	1	2,69	2,69	1,74	9,26
Error	7	6,67	0,95	-	-
Total	11	18,78	-	-	-

Based on the data processing above, it is found that the factors that influence the quality of coffee pulp screen printing ink are coffee grounds (A) and Water (B). The quality characteristics used are Nominal is The best, then the selection of factor levels is based on the largest level value, namely the AxB factor level 1. so that it produces an optimal combination of settings A1, D1, AxC1, AxD1 and AxB1.

Table 11. Selected Combination Results

Code	Factors	Level
A	Amount of Coffe Pulp	0,5 Kg
B	Dose Of Water	0,5 Liter
C	Amount Of Vinegar	100 ml
D	Cooking Time	10 Menit

Table 12. Confirmation trial results

Experiment	Confirmation
1	9
2	10
3	10
4	9
5	10
6	9
7	10
8	10

From table 12, the assessment of whether or not the confirmation experimental results are accepted is done by comparing the confidence interval between the predicted results of the response under optimal conditions and the results of the confirmation experiment. Optimal prediction confidence interval obtained at 30.07 so that the optimal predictive confidence interval is 28.35 ≤ 30.07 ≤ 31.19. Confirmation interval was obtained at 22.27 so that the interval is between 20.40≤ μ_predictions ≤ 24.14.

4. Conclusion

The conclusion of this research are:

1. After conducting research and experiments conducted on the floor of coffee pulp screen printing production on product quality, several conclusions can be drawn. Factors that significantly influence the quality of paving products among others are: Coffee Pulp factor by 1.49%, interaction of Coffee Pulp factor and Water Factor and Cooking Time factor by 9.26%.
2. The optimal combination of settings to reduce the number of defects and improve the quality caused by process factors and raw materials are 0.5 kg Coffee Pulp and Cooking Time 15 Minutes, then the optimum interaction level combination is 0.5 kg Coffee Pulp and 0.5 Liter Water, 0.5 kg Coffee Pulp and 10 Minute Cooking Time , 0.5 kg of coffee grounds and 100 ml vinegar. S so that it can be concluded that the optimal combination obtained can improve quality by reducing the level of disability that occurs.

5. References

[1] Wibowo and Praptiningsih 2013 Management And Development Of Garment Business In Indo *Surabaya Garment Business Management vol 1*

[2] Soejanto I 2009 Design Of Experiment Using Taguchi Method *Graha Ilmu* **292**
[3] Gunawan B 2015 Taguchi Method As One Alternative To Quality Control *Journal of Accounting and Investment vol 2* **45-55**
[4] Anggraini D, Dewi S K and Saputro T E 2015 Application of the Taguchi Method for Lowering Levels Defects In Paving Products *Journal of Industrial Engineering vol 16* **1-9**
[5] Pundir R, Chary G H V C, and Dastidar M. G. 2018 Application of Taguchi method for optimizing the process parameters for the removal of copper and nickel by growing *Water Resources and Industry 20* **83–92**
[6] Hartono M Improve the quality of plastic products with the Taguchi method *Journal of Industrial Engineering*
[7] Pawar K and Palhade R 2015 Multi-objective Optimization of CNC Turning Process Parameters for High Speed Steel (M2) Using Taguchi and ANOVA Method *International Journal of Hybrid Information Technology vol 8* **67-80**
[8] Chung Y T, Ba-Abbad M M, Mohammad A W, Hairom N H H, and Benamor A 2015 Synthesis of minimal-size ZnO nanoparticles through sol–gel method: Taguchi design optimisation *Materials & Design vol. 87* **780-787**

IOP Conf. Series: Materials Science and Engineering **931** (2020) 012024 doi:10.1088/1757-899X/931/1/012024

Analysis of of Consumer Preferences on Aceh Songket Product Using Conjoint Analysis (Case of Study: Banda Aceh City and Aceh Besar District)

Ilyas[1], S Zuhri[1], E Yusra[1], CM Maydini[1], F Erwan[1]
[1]Industrial Engineering Department, Universitas Syiah Kuala, Darussalam, Banda Aceh, Indonesia

E-mail: ilyasishak@unsyiah.ac.id

Abstract. Indonesia as the country rich in cultural heritage, has a variety of characteristics including woven fabric. Songket fabric is one form of woven fabric that is owned by almost every region. Aceh is one of the best producing areas for songket because of its quality. However, the limitations of the songket design made songket fabric not develop as the expectation. Therefore, the innovation is needed based on the consumer preferences by combining songket with other products. This research aims to identify the consumer preferences toward the product. The method that was used to analyze consumer preferences was conjoint analysis. The results showed that there were five attributes that influenced consumers in choosing songket products based on the highest level of importance for each attributes which were basic color at 35,381%, type of motive at 26,650%, type of product at 16,195%, materials at 11,718% and motive color at 10,057% while the most preferred combination product based on the highest utility level results from the level of each attributes were clothing product at 0,089, with black as basic color at 0,110, gold color as color motive at 0,012, using polyester and embroidery as materials at 0,059 and pinto aceh as a motive 0,089.

1. Introduction

As a country which has a long history in trading, Indonesia is a country with a strategic region in Southeast Asia. With this trade relationship, a craft of art in a number of regions in Indonesia is created, especially weaving [1].

Traditional woven cloth is a masterpiece which has been passed down for generations made by weaving weft and warp threads on a loom. Woven cloth consists of several types including, double tie weaving, weft weaving, warp weaving, simple weaving and songket weaving. Songket woven fabric is a woven fabric using a non-machine loom [2]. Songket is a famous type of woven fabric in Indonesia, although there are differences in the motives and ways of weaving between one region and another [3].

Aceh is one of the regions which has a special craft in the form of woven cloth, namely songket fabric. The difference between Aceh songket and other regions is the tight texture of the thread, namely bounded warp threads instead of the weft thread, as in songket fabric from other regions [4].

Until now, the market share for songket is quite extensive and developing in the local and national regions which are in demand by consumers from the middle to upper levels. Songket products on the market are still in the form of fabric while consumers want songket fabric combined with products which can be used daily, producing a variety of products. This phenomenon makes songket requires a touch of innovation in accordance with consumer needs that can be known through consumer desires.

In this case, the method that can be used to obtain consumer preferences is conjoint analysis. Conjoint Analysis produces a design of the product which most consumers want. Conjoint analysis aims to determine consumer perceptions of objects consisting of several parts [5]. Based on the background, the research was conducted on the analysis of consumer preferences for songket Aceh products with a conjoint analysis approach to help consumers in realizing their preferences for interested products.

2. Methods

This research was conducted around Banda Aceh city and Aceh Besar district. Researcher visited public places, community gathering and songket gallery. In the preliminary stage, problem identification was carried out by conducting a literature study and field study then formulating the problem and determining the objectives and benefits of the research.

Data collection was done by determining the initial factors and levels. There are 6 factors with each level, namely the product factor consisting of 6 levels, namely clothes, bags, wallets, sandals, caps and souvenirs. Then the motive factor consists of 2 levels, namely plain and mixed. Furthermore the basic color factor consists of 10 levels namely blue, black, brown, pink, green, red, orange, yellow, gray and purple. For the color factor motive consists of 3 levels, namely gold, silver and white. The material factor consists of 5 levels, namely embroidery polyester, rayon polyester, cotton embroidery, cotton rayon and silk. Finally, the last factor is the type of Aceh unique motive consisting of 15 levels, namely *akie mira pati*, *awan meutalo*, *bukulah*, *bungong campi*, *bungong kupula*, *bungong mulu*, *bungong pucok reubong*, *halua reuteuk*, *Laailahailallallah* sentence, *kertah canden*, *pinto aceh*, *pucok aron*, *pucok geulima*, *pucok meuriya* and *reubong aceh* (*utom*). Furthermore, the first stage questionnaire was made using the initial data, then the 50 people of respondents were determined. This first stage questionnaire was distributed to the 50 respondents and finally the questionnaires were collected for data recapitulation.

The recapitulated data was then processed using SPSS software using orthogonal arrays commands to produce a combination of factors and levels. Afterwards, the second stage questionnaire containing a combination of factors and levels selected was designed then continued by distributing it to the same respondents as the first stage. The results of the second phase questionnaire were collected to be recapitulated. The validation and reliable tests were conducted to the collected data to obtain valid and reliable data. If it invalid, then the questionnaire would be redistributed and if it was valid and reliable, it could be proceeded to data processing with conjoint analysis using SPSS software. Data processing was using the syntax command by using orthogonal design data and correspondent preference data.

3. Result and discussion

Determination of factors and initial levels is needed to obtain a combination of selected factors and levels which will be used as a stimulus card to be assessed by respondents through the second stage questionnaire. The selected factors and levels were determined based on the highest frequency results from the recapitulation results of all respondents in the first stage questionnaire that was up to a frequency limit of 5. The following is a summary of the factors and levels selected from the songket product.

Table 1. Songket Product's Factors and Levels

No.	Factor	Level
1.	Product	Clothes
		Bag
2.	Basic Color	Black

		Red
		Purple
		Red Yellow Green
		Black Pink
3.	Motive Color	Gold
		Silver
4.	Material	Polyester Embroidery
		Silk
5.	Motive Type	*Pinto Aceh*
		Bungong Campi
		Bungong Meulu
		Pucok Aron

Next to get a stimulation card combination in advance was to determine the presentation method. In the presentation method, the conjoint methodology used was traditional conjoint and the basic form model used was the additive model. The type of presentation method used was full profile because it is possible to reduce the number of factor comparisons using fractional factorial design.

Obtaining the results of this stimulation was done using SPSS software using the help of orthogonal arrays commands and resulting the orthoplan results which were 25 stimuli. Next is the display of stimuli cards (plan cards).

No.	Card ID	Product Type	Basic Color	Motive Color	Material	Motive Type
1	1	Clothes	Red	Silver	Polyester and Embroidery	*Bungong Campi*
2	2	Clothes	Purple	Gold	Polyester and Embroidery	*Bungong Campi*
3	3	Bag	Black Pink	Gold	Polyester and Embroidery	*Pinto Aceh*
4	4	Bag	Red	Gold	Polyester and Embroidery	*Pinto Aceh*
5	5	Clothes	Black	Gold	Silk	*Bungong Meulu*
6	6	Bag	Purple	Silver	Silk	*Pucok Aron*
7	7	Bag	Purple	Gold	Silk	*Pinto Aceh*
8	8	Bag	Red	Gold	Silk	*Bungong Meulu*
9	9	Clothes	Black	Silver	Silk	*Pinto Aceh*
10	10	Clothes	Red	Gold	Polyester and Embroidery	*Pucok Aron*
11	11	Clothes	Red Yellow Green	Gold	Silk	*Pucok Aron*
12	12	Clothes	Red	Silver	Silk	*Pinto Aceh*
13	13	Clothes	Red Yellow Green	Gold	Polyester and Embroidery	*Bungong Meulu*
14	14	Bag	Black	Gold	Polyester and Embroidery	*Bungong Campi*
15	15	Bag	Black	Silver	Polyester and Embroidery	*Pucok Aron*
16	16	Bag	Black Pink	Silver	Polyester and Embroidery	*Bungong Meulu*
17	17	Clothes	Black Pink	Silver	Silk	*Bungong Campi*
18	18	Clothes	Black	Gold	Polyester and Embroidery	*Pinto Aceh*
19	19	Clothes	Purple	Silver	Polyester and Embroidery	*Bungong Meulu*
20	20	Bag	Red Yellow Green	Gold	Silk	*Bungong Campi*
21	21	Bag	Red Yellow Green	Silver	Polyester and Embroidery	*Pinto Aceh*
22	22	Clothes	Black Pink	Gold	Silk	*Pinto Aceh*
23	23	Clothes	Red Yellow Green	Silver	Polyester and Embroidery	*Pinto Aceh*
24	24	Clothes	Purple	Gold	Polyester and Embroidery	*Pinto Aceh*
25	25	Clothes	Black Pink	Gold	Polyester and Embroidery	*Pucok Aron*

Figure 1. Stimulation Card.

The orthogonal design data was obtained from the results of the first stage questionnaire data processing, namely the data that showed a combination of factors and levels. This data was needed to be used as data contained in data processing syntax. Preference data was obtained from the results of respondents' answers on the second stage questionnaire. The data was in the form of rating choices on 25 stimuli cards (showed in Figure 1) by 50 respondents. This data was also needed to be used as data contained in data processing syntax. The syntax page on the SPSS application provides data processing results, which contain the level of importance and the best combination of the second stage questionnaire. Following are the results of data processing syntax.

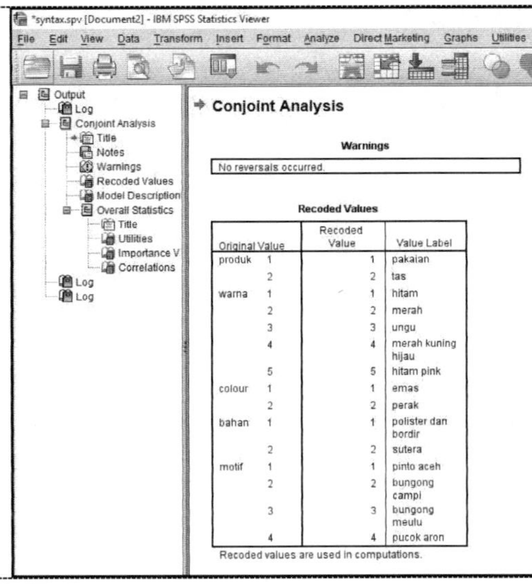

Figure 2. Recorded Value.

Figure 2 shows the number of factors and levels as well as the details of Songket products. The factors are product with clothes and bag as levels, basic color with black, red, purple and black pink as levels, motive color with gold and silver as levels, material with polyester and embroidery as levels and motive type with *Pinto Aceh*, *Bungong Campi*, *Bungong Meulu*, *Pucok Aron* as levels in sequence.

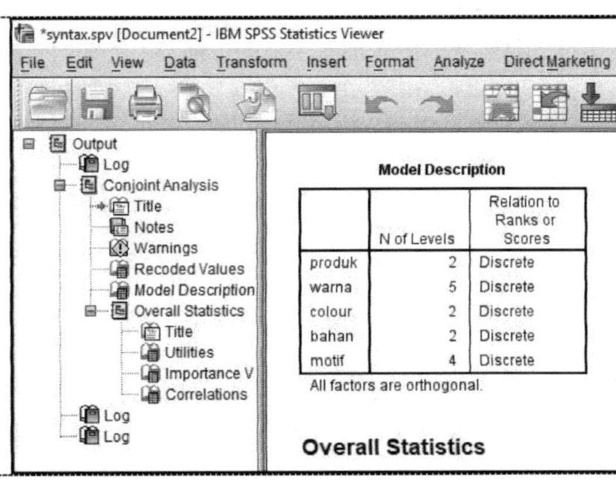

Figure 3. Factor Model Description.

Figure 3 shows a description of the model of discrete value factors (product, basic color, motive color, material and motive type) meaning that the data was a qualitative data.

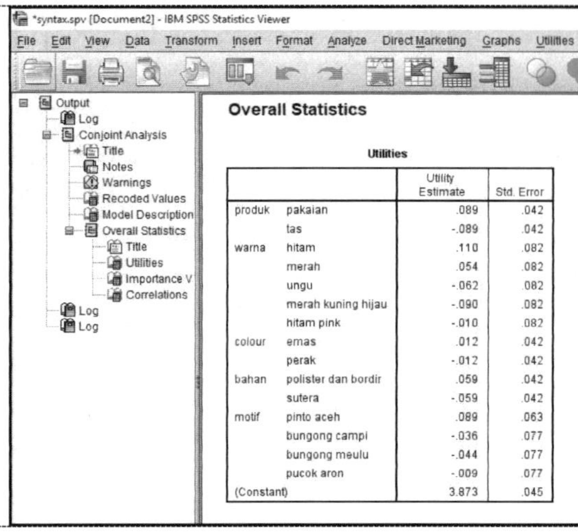

Overall Statistics

Utilities

		Utility Estimate	Std. Error
produk	pakaian	.089	.042
	tas	-.089	.042
warna	hitam	.110	.082
	merah	.054	.082
	ungu	-.062	.082
	merah kuning hijau	-.090	.082
	hitam pink	-.010	.082
colour	emas	.012	.042
	perak	-.012	.042
bahan	polister dan bordir	.059	.042
	sutera	-.059	.042
motif	pinto aceh	.089	.063
	bungong campi	-.036	.077
	bungong meulu	-.044	.077
	pucok aron	-.009	.077
(Constant)		3.873	.045

Figure 4. Utility Level.

Figure 4 shows the utility level of each level. The most preferred level is the highest positive value.

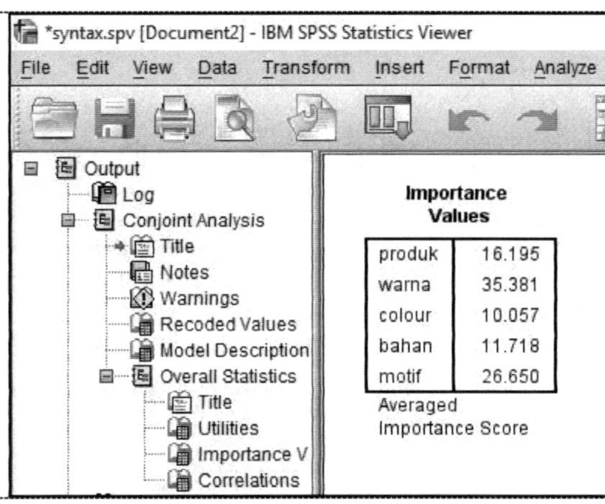

Importance Values

produk	16.195
warna	35.381
colour	10.057
bahan	11.718
motif	26.650

Averaged Importance Score

Figure 5. Importance Level.

Figure 5 shows the percentage of importance of the factors that most consumers pay attention in choosing songket products. It shows that color is the most preference factor for respondents in choosing songket product.

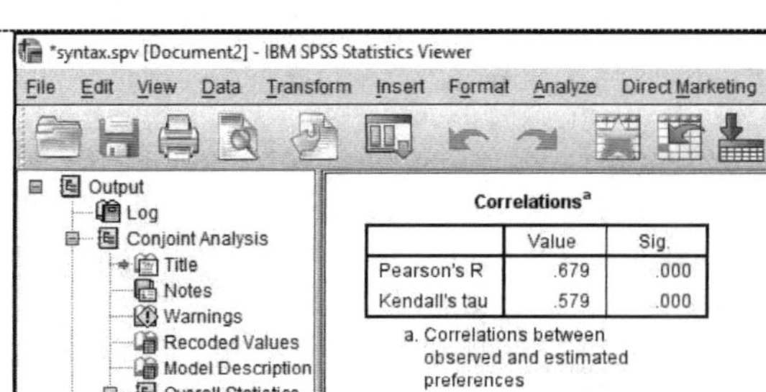

Correlations[a]

	Value	Sig.
Pearson's R	.679	.000
Kendall's tau	.579	.000

a. Correlations between observed and estimated preferences

Figure 6. Research Correlation.

Figure 6 shows the correlation that the model obtained was accurate so the results of the data calculation were worthy of further analysis. This shows that the results of the model obtained were not much different from the actual opinions of respondents.

Based on the acquisition of utility level results, it could be known that the combination of the level factor of the songket product that was most preferred by consumers, was the highest level of utility of each level on each factor, then the combination of 18 cards which was the most desirable consumer were selected, namely clothing products with a black basic color, the color of the gold motive, made from polyester and embroidery and using aceh pinto motive. The following is an illustration of the preferences of selected consumers.

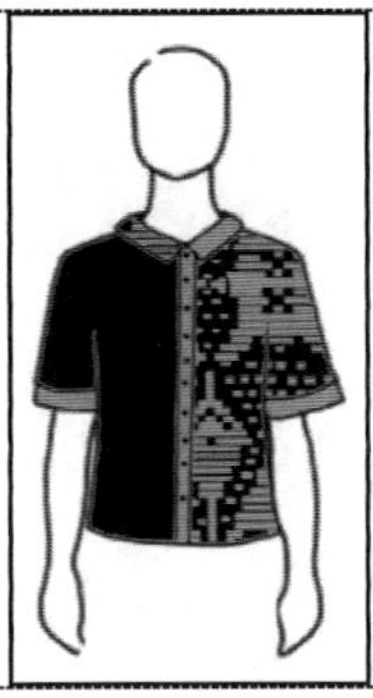

Figure 7. The Chosen Songket Product.

To discover the assessment of other combinations can be done by adding the value of each level contained at the utility level. The following is an assessment of consumer preferences for songket combination products from the most liked to the least preferred.

Table 2. Value and Ranking from 25 Stimuli Cards

Card	Value	Ranking
1	0.154	6
2	0.062	12
3	0.061	13
4	0.125	9
5	0.108	11
6	-0.231	24
7	-0.109	22
8	-0.126	23
9	0.217	2
10	0.205	3
11	-0.057	20
12	0.161	5
13	0.026	17
14	0.056	15
15	0.059	14
16	-0.096	21
17	-0.028	18
18	**0.359**	**1**

19	0.030	16
20	-0.262	25
21	-0.043	19
22	0.121	10
23	0.135	8
24	0.187	4
25	0.141	7

To determine the consumer preferences against factors which are most desirable, can be seen based on the acquisition of importance, namely the level of importance. The following is a graph of the acquisition of the importance value of factors in songket products.

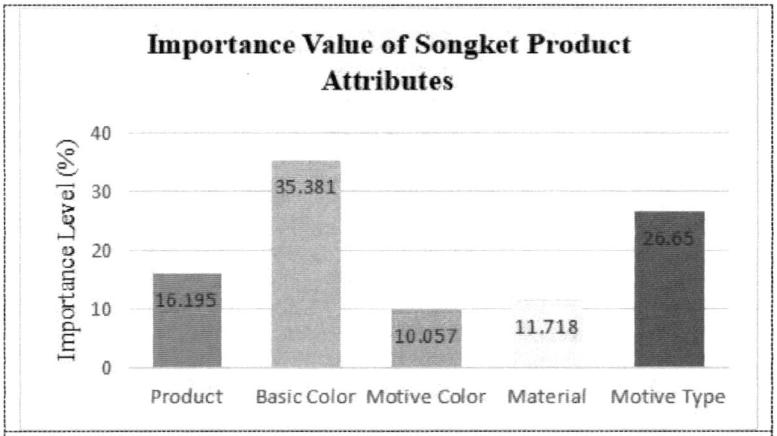

Figure 8. Graphic of Factors Importance Level.

Based on Figure 8, the consumer preference results were obtained that the factor which most influences consumers in choosing songket products were the basic color then the type of motive, the type of product, the material and finally the color of the motive.

4. Conclusion

The conclusions of this research are:

a. Factors and levels selected based on the results of consumer preferences for Aceh songket products were factors of product type consisting of clothing and bag levels; basic color factors with levels of black, red, purple, red yellow green and black pink; color factor motives consisting of gold and silver color levels; material factors consisting of embroidery polyester and silk levels and factors of the type of motive with the levels of aceh pinto, bungong campi, bungong meulu, and aron pucok motives.

b. Factors that influence consumers in choosing songket products based on their level of importance were the basic color factor with the highest percentage of other factors in the amount of 35.381%. Then the next sequence was the type of motive at 26,650%, the type of product at 16,195%, songket material at 11,718% and the last was the color of the motive with the lowest percentage at 10,057%.

c. The best combination was obtained based on the highest utility value at the level of each factor so that 18 cards were selected to be the most preferred combination of consumers namely the clothing product type with black basic colors, gold motive colors, using polyester and embroidery material and finally with aceh pinto motive. The utility levels of each level were 0,089 for the clothing level, 0,110 for the black color level, 0,012 for the gold motive color level, 0,059 for the polyester and embroidery material level and 0,089 for the aceh pinto motive level.

5. References

[1] Viatra A W and Triyanto S 2014 Seni Kerajinan Songket Kampoeng Tenun Di Indralaya *Ekspresi Seni 16 November (2)* **168–183**

[2] Marianti M M and Istiharini 2014 Analisis Karakteristik dan Perilaku Konsumen Tenun Songket Palembang

[3] Purwanto R and Siregar S M 2016 Sejarah Songket Berdasarkan Data Arkeologi *Siddhayatra 21(2)* **97–106**

[4] Leigh B 1989 Tangan-Tangan Trampil, Seni Kerajinan Aceh (Hands of Time, the Crafts of Aceh) *Djambatan*

[5] Santoso S 2010 Statistik Multivariat Konsep dan Aplikasi dengan SPSS *PT Elex Media Komputindo*

Burnout Syndrome analysis among hospital nurses using Maslach Burnout Inventory - Human Service Survey (MBI-HSS): a case study

F Erwan[1,3], M Iqbal[2], I Hasanuddin[1,3], S Zuhri[1], C M Maydini[1]
[1]Industrial Engineering Department, Universitas Syiah Kuala, Darussalam, Banda Aceh, Indonesia
[2]Department of Mechanical and Industrial Engineering, Universitas Syiah Kuala, Darussalam, Banda Aceh, Indonesia
[3]Work System Design and Ergonomics Laboratory, Industrial Engineering Study Program, Universitas Syiah Kuala, Darussalam, Banda Aceh, Indonesia

E-mail: friesca_erwan@unsyiah.ac.id

Abstract. Burnout is a psychology syndrome which experienced by any professions throughout their careers. This study analyses burnout syndrome encountered by hospitals' nurses based on the Maslach Burnout Inventory-Human Service Survey (MBI-HSS) and provides findings by looking at the dimension of burnout and demographic factors which consist of sex, age, marital status, level of education, years of career, and duty station. This study reveals that; the overall value of burnout experienced by hospitals' nurses is categorized into level 1 with the highest value presented by dimension 1 (physical exhaustion), and burnout generally experienced by man, nurses younger than 30 years old, married nurses, lower degree nurses, nurses whose less than 20 years in career, and nurses whose responsible in terminal care and emergency unit. In addition, data obtained from the survey is utilized to determine burnout causative factors using fishbone and pareto diagram.

1. Introduction

Burnout is a psychology syndrome to define a situation in which someone feels extremely exhausted emotionally, experience depersonalization and low personal accomplishment. The term was first introduced by a psychiatrist named Herbert Freundenberger following his observation on young and idealistically motivated volunteers in a trauma research centre for drug addicts. Freundenberger employed the term of burnout to explain a situation in which someone experienced "a gradual energy depletion and loss of motivation and commitment, accompanied by a wide array of mental and physical symptoms." At about the same time, a social psychological researcher named Christina Maslach observed health care professionals in coping with their emotional arousal at work and revealed that many of them experienced emotional exhaustion and a crisis in their professional competence as well as developed negative perceptions towards their patients, in which Maslach defined as burnout.

Burnout experienced by nurses shows a significant value, the highest and exceed 40% among other health care professionals, particularly those working in the Intensive Care Unit (ICU), emergency, or terminal care (UGD). To analyse and assess burnout that occurs, Maslach and Jackson developed a Maslach Burnout Inventory-Human Service Survey (MBI-HSS) in 1981, focusing on health and community service [2]. The MBI-HSS is the most popular and effective measurement tool used for measuring burnout in the health profession and has been validated through extensive research conducted more than 25 years since its introduction. MBI-HSS has been adopted to analyse burnout that occurs in many professions such as teacher, factory labour, librarian, employees at profit and non-profit organizations, and also working students. MBI-HSS has also been collaborated with other methods to obtain relevant analysis based on the chosen case studies. For example, [3] has conducted a study of burnout among nurses using Work Characteristic Survey and Nordic Musculoskeletal Symptom Survey, and revealed that work-related factors have a contribution to musculoskeletal problems experienced by nurses and potentially lead to burnout.

This study aims at determining the level of burnout experienced by hospitals' nurses in Banda Aceh. Nurses are often faced with non-nursing tasks such as completing patient administration, sending prescriptions and taking drugs or medical devices to the pharmaceutical department, taking patients to radiology and laboratories, accompanying doctor visits to patients, and others. Although additional staff members have been provided, non-nursing tasks remain a workload that cannot be avoided by nurses. These excessive workloads may have a negative influence on workers [4].

2. Research method

This study employs MBI-HSS to obtain information about burnout that occurs among hospitals nurses in Banda Aceh. The MBI-HSS contains 22 items, which are divided into three dimensions of burnout namely emotional exhaustion, depersonalization and low personal accomplishment. Table 1 shows the MBI-HSS developed by Maslach in determining burnout levels among services workers. Furthermore, the MBI-HSS was distributed to 100 nurses in three public hospitals in Banda Aceh: RSUDZA (Rumah Sakit Umum Daerah Zainal Abidin), BLUD RSIA (Balai Layanan Umum Daerah Rumah Sakit Ibu dan Anak) and RSUD Meuraxa (Rumah Sakit Umum Daerah Meuraxa).

Table 1. Burnout factors

Burnout dimension	Variable No.	Burnout factors
Emotional exhaustion (dimension 1)	1	I feel emotionally drained from my work.
	2	I feel used up at the end of the workday.
	3	I feel fatigued when I get up in the morning and have to face another day on the job.
	4	I feel I'm working too hard on my job.
	5	I feel like my work is breaking me down.
	6	I experience musculoskeletal problems after finishing my work.
Depersonalization (dimension 2)	7	Working with people all day is really a strain for me.
	8	I feel burned out from my work.
	9	I fell frustrated by my job.
	10	I feel I'm at the end of my rope.
	11	I feel I treat some recipients as if they were impersonal objects.
	12	I have become more insensitive to people since I've been working.
	13	I feel recipients blame me for some of their problems.
	14	I don't really care what happens to some recipients.

IOP Conf. Series: Materials Science and Engineering **931** (2020) 012025 doi:10.1088/1757-899X/931/1/012025

Table 1. Burnout factors (cont.)

Burnout dimension	Variable No.	Burnout factors
Low personal accomplishment (dimension 3)	15	I cannot deal with emotional problems very calmly
	16	I have not accomplished may worthwhile things in this job.
	17	I feel unenergetic at work.
	18	I fell I don't have achievements at work.
	19	I feel I'm influencing other people's live negatively through my work
	20	I have difficulties in creating a relaxed atmosphere with my recipients.
	21	I couldn't look after my patients' problems very effectively.
	22	I feel I don't have enough energy to handle my recipients.

In addition, this study adds demographic information section at the beginning of the survey to obtain respondents' demographic data such as sex, age, marital status, level of education, years of career, and duty station. The information will be used to relate demographic factors with burnout that occur among nurses. The overall data obtained from the MBI-HSS will determine the average value of burnout based on its three dimensions, dominant factors which generate burnout and the average value of burnout based on the demographic factors.

3. Result and discussion

The first results of MBI-HSS is the average value of burnout based on its three dimensions which shown in Figure 2. Dimension 1 has the highest average value of burnout (as of 1.98) while the lowest average value represented by dimension 2 (1.34). Based on the stage of burnout syndrome stated by [5], the average value obtained is categorized in to the first stage with range between 1 to 2. The first stage of burnout syndrome indicates a person (nurse) still has high hopes and idealism towards work which is characterized by high enthusiasm for work, shows dedication and commitment to work, shows high energy and achievement, is constructive and positive, and has good views [5].

In addition, the average value of burnout obtained is influenced by the average answers of respondents who tend to choose scale 1 (never) and scale 2 (rarely) on the MBI-HSS. The average value of burnout obtained also represents that burnout experienced by nurses is still within the limits which can be mitigated. However, in details, the highest average value of burnout which represented by dimension 1 closes to stage 2 of burnout. Thus, it needs attention to minimize the level of burnout. Stage 2 of burnout is indicated by a person (nurse) shows symptoms of initial stress and starts to be pessimistic and dissatisfied with their work, characterized by experiencing physical and mental fatigue, feels frustrated and filled with illusions, decreasing in work morale, and experiencing boredom [5].

The second result of MBI-HSS is burnout dominant factors, calculated using factor analysis. This study reveals that there are ten factors which significantly contribute to burnout that occurs amongst hospital nurses. Table 2 shows the burnout dominant factors in which 1 factor refers to dimension 1, 4 factors refer to dimension 2, and 5 factors refer to dimension 3. The highest factor value is represented by the burnout factor of "*I have difficulties in creating a relaxed atmosphere with patients*," with a value of 0.920. Meanwhile, the lowest value is represented by the factor of "*I feel I treat some recipients as if they were impersonal objects*," with a value of 0.802. The highest and lowest factors are related to the condition of nurses in serving the patients. There is a tendency that high workload enforces nurses to have ideal time in providing services to patients; thus, consultation needed by patients becomes a nuisance for nurses.

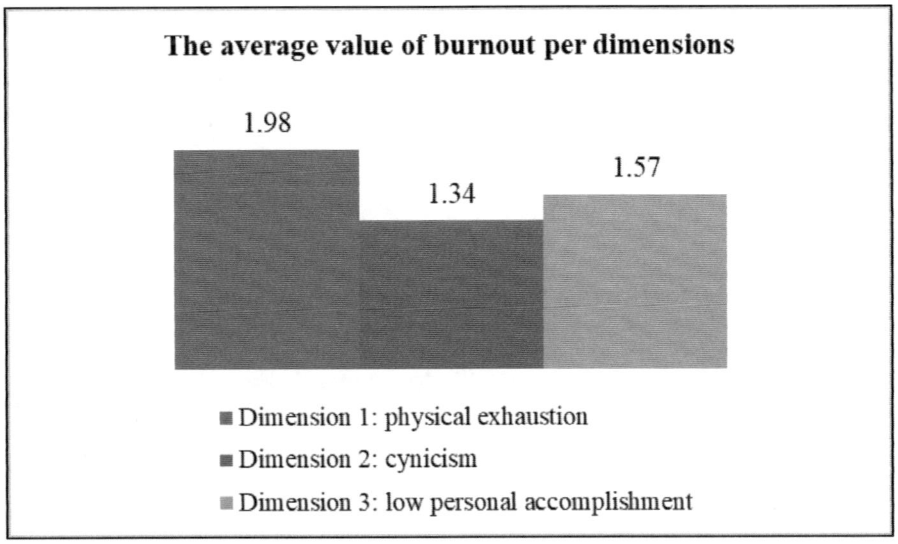

Figure 2. The average value of burnout per dimension

Table 2. Burnout dominant factors

Burnout dimension	Variable No.	Burnout dominant factors	Value
Dimension 1	1	I feel used up at the end of the workday.	0.858
Dimension 2	2	I feel burned out from my work.	0.838
	3	I feel I treat some recipients as if they were impersonal objects.	0.802
	4	I feel recipients blame me for some of their problems.	0.880
	5	I don't really care what happens to some recipients.	0.884
Dimension 3	6	I have not accomplished may worthwhile things in this job.	0.813
	7	I feel I'm influencing other people's live negatively through my work.	0.833
	8	I have difficulties in creating a relaxed atmosphere with my recipients.	0.920
	9	I couldn't look after my patients' problems very effectively.	0.875
	10	I feel I don't have enough energy to handle my recipients.	0.862

The third results of MBI-HSS is the level of burnout based on the demographic factors, which are sex, age, marital status, level of education, years of career, and duty station. Based on sex factor, Figure 3 shows that burnout experienced more on male nurses compared to female nurses. The condition is influenced by male internal factors such as characteristic [6]. In general, men have less emotional nature in social life and prefer to hide their feelings to public when facing a problem. Such conditions are very vulnerable to stress and may lead to burnout. Whereas, women experiencing the opposite condition and tend to involve emotionally in social life. Thus, when experiencing problems or obstacles at work, women have better ability to convey their feelings to others (for example to co-workers) and suppress excessive emotion that may lead to burnout.

Based on the age factor, Figure 4 shows that burnout was experienced more by nurses under the age of 30 compared to those with the age of 30 or older. The condition is supported by [7], i.e. the level of burnout experienced by young workers (in this case is nurses under the age of 30) is higher compared to workers aged 30-40 years. According to [1], young age is synonymous with a lack of work experience due to reality shock and an indication of an identity crisis due to unsuccessful occupational socialization. Hence, when facing problems at work, young nurses tend to feel scared and depressed of making mistakes that make their emotions uncontrollable and vulnerable, and this may lead to stress and burnout.

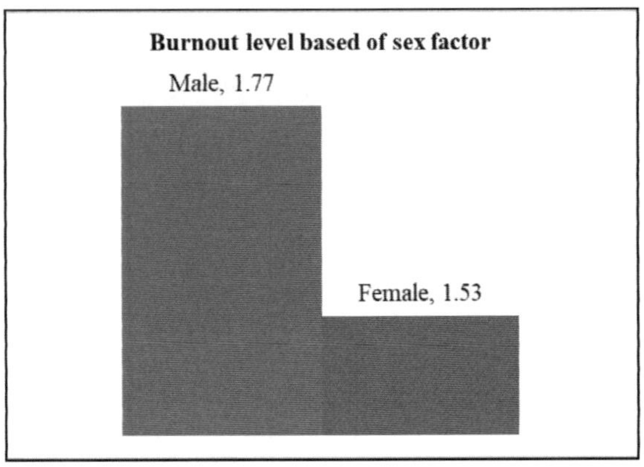

Figure 3. Burnout level based on sex factor

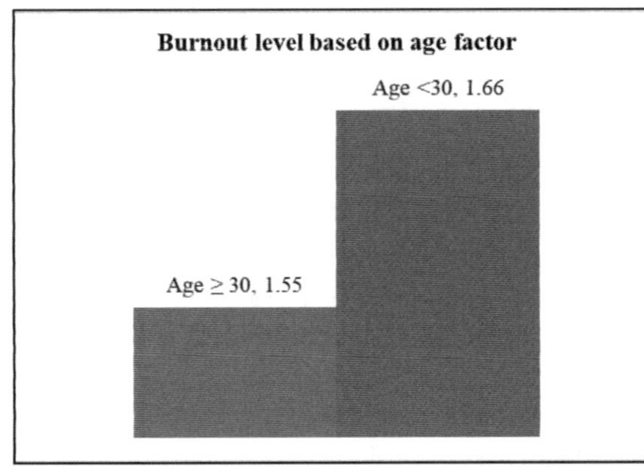

Figure. 4. Burnout level based on age factor

Based on marital status, Figure 5 shows that burnout was experienced more by married nurses compared to single nurses. The results of this study imply that burnout experienced by married nurses occurred due to multiple workloads faced both at work and also at home. However, previous studies revealed that unmarried (single) nurses are more susceptible to burnout [8, 9, 10, 11] due to a lack of role and social support from the family which may help someone in overcoming emotional demands at work [6]. Also, people who have a family tend to have a more realistic view [12]. Thus they will seek support and opinions from their family members whenever they face problems at work, which may reduce the level of stress.

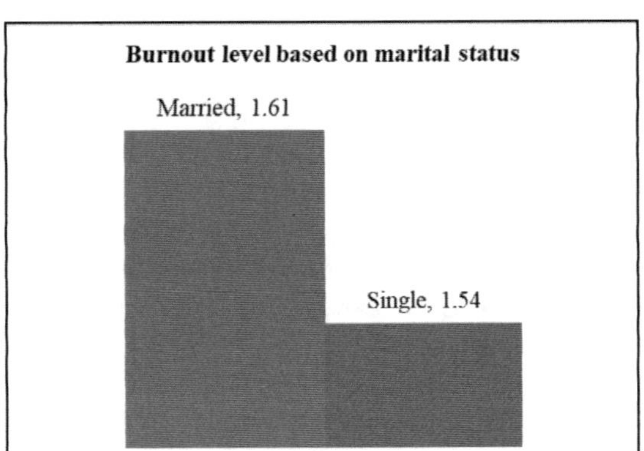

Figure 5. Burnout level based on marital status

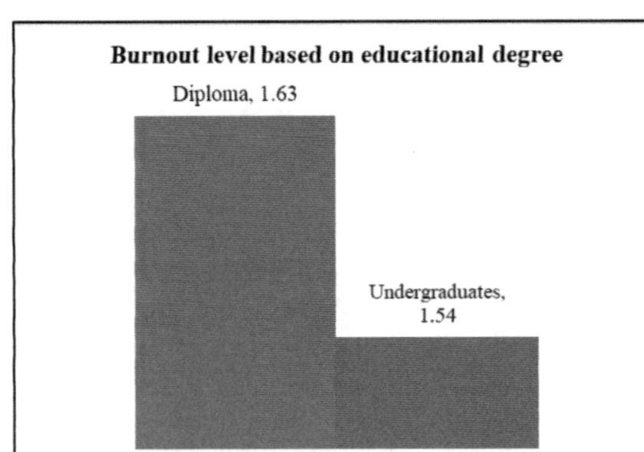

Figure 6. Burnout level based on educational degree

Based on the level of education, Figure 6 shows that burnout experienced more on diploma nurses compared to bachelor nurses. The results of this study imply that diploma nurses have more technical tasks, as a consequence, their workloads are higher than the bachelor nurses. In addition, diploma nurses serve as implementer while bachelor or higher degree nurses serve as the head of department or unit which enables diploma nurses to experience demanding and stressful job. However, the results of this study contradict from the literature and previous studies, which revealed that burnout is more vulnerable to be experienced by a professional or highly educated because of the high expectations for the quality of work to be achieved.

Based on work experience, Figure 7 shows that burnout was experienced more by nurses with less than 20 years in service compared to nurses who have been in service longer. The results of this study are supported by [7] who mentioned that burnout is vulnerable at the beginning of a career due to a lack of work experience. However, a study revealed that burnout is also experienced by senior workers due to a long period of work which leads to physical and emotional burnout [13].

Finally, based on nurses' duty station, Figure 8 shows that burnout was experienced more by nurses responsible at the emergency department. This condition is consistent with the literature and previous studies, in which emergency or intensive care units nurses usually have higher workload with dynamic and fluctuating work conditions compared to nurses who served in polyclinics or inpatients room.

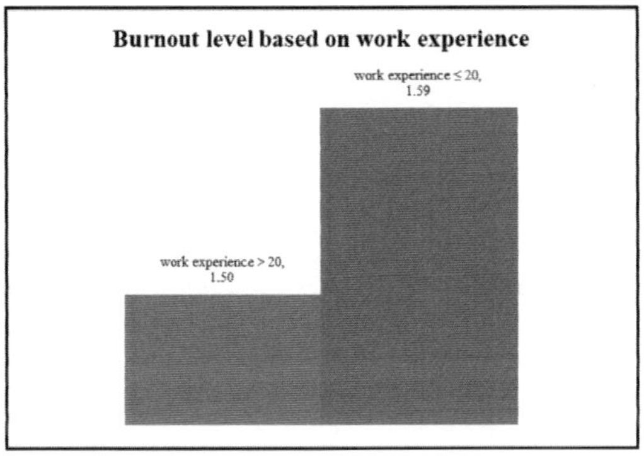

Figure 7. Burnout level base on work experience

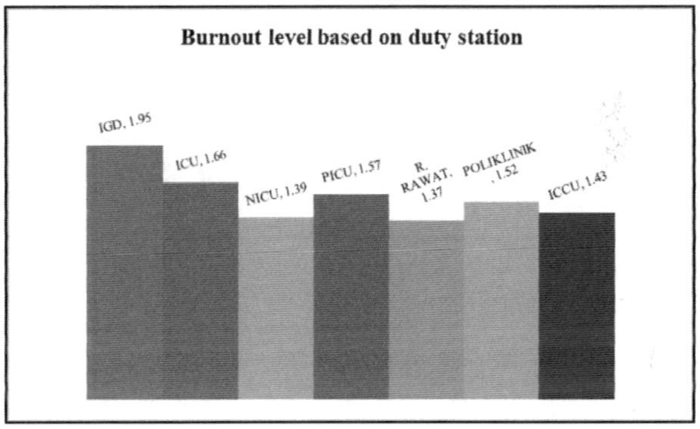

Figure 8. Burnout level based on the duty station

Furthermore, burnout dominant factors obtained previously is visualized using a fishbone diagram (Figure 9) to show the cause and effect of burnout that occurred among hospital nurses. The contributing factors were then sorted from the highest to the lowest to provide a priority list of factors that need to be solved immediately. The list is visualized into a pareto diagram shown in Figure 10. The tallest bar on the chart implies the most serious problems that occur among nurses, which contribute to burnout, while the lowest one implies the least problematic factor. Based on the pareto concept 80/20 (80% problems occur as a result of 20% cause), there are two priority factors which refer to 20% cause in nursing systems and need to be solved immediately to minimize the level of burnout experienced by nurses.

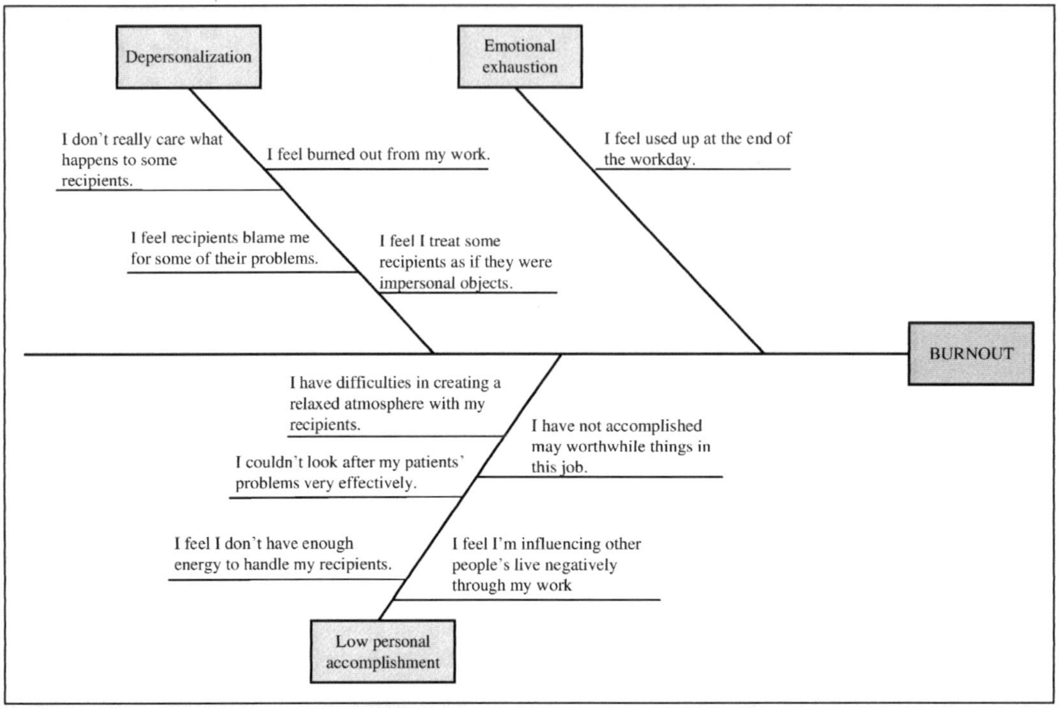

Figure 9. Fishbone diagram of burnout

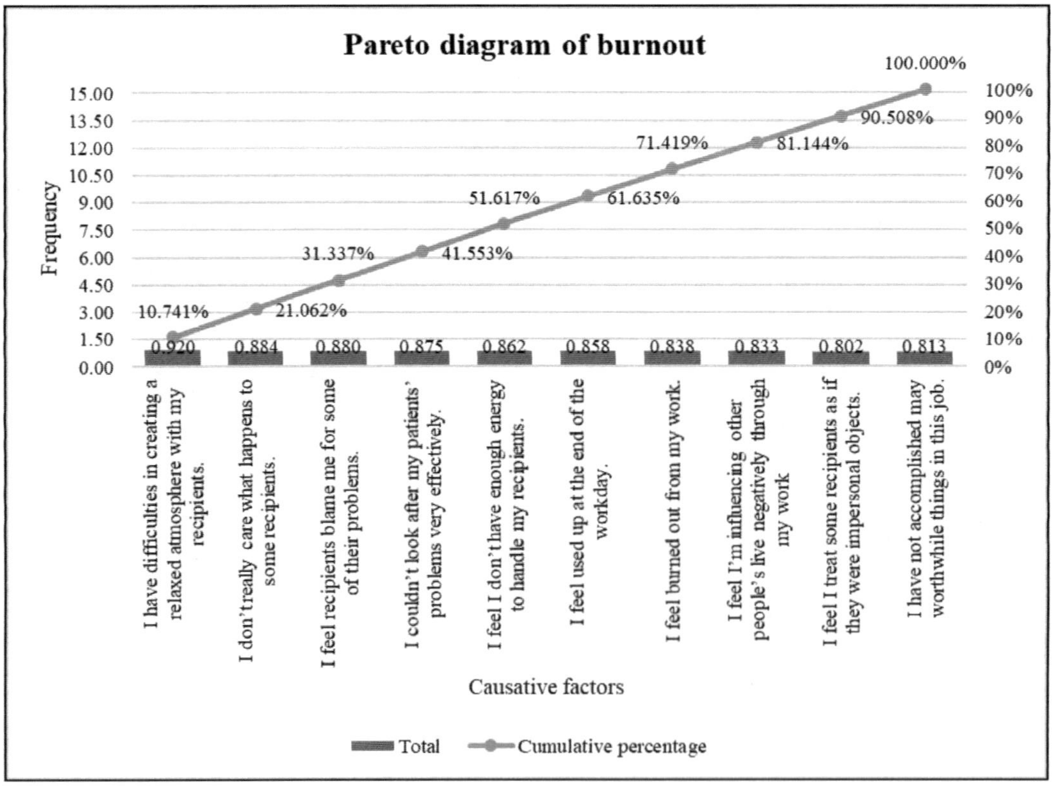

Figure 10. Pareto diagram of burnout

IOP Conf. Series: Materials Science and Engineering **931** (2020) 012025 doi:10.1088/1757-899X/931/1/012025

4. Summary

To conclude, based on MBI-HSS, the average value of burnout refers to level 1 which means the condition of burnout experienced by hospital nurses is still within limits and can be quelled. The condition also indicates that nurses still have dedication for work characterized by enthusiasm and commitment to work, energy and achievement, they are constructive and positive, and they have positive perception. In addition, from 22 burnout items in the MBI-HSS, 10 items were selected as dominant factors which contribute to burnout that occurs among hospitals' nurses and serve as the contributing factors in the fishbone diagram. The contributing factors were then analysed using pareto diagram to provide a priority list of factors which need to be solved or need immediate actions to minimize burnout among nurses. Furthermore, based on the demographic aspect, MBI-HSS reveals that burnout generally experienced by male nurses under the age of 30, married nurses, lower degree nurses, nurses who are less than 20 years in career, and nurses who are responsible for terminal care and emergency unit.

5. Reference

[1] Schaufeli W B 2007 Burnout in Health Care in: Pascale Carayon (Ed.) *Handbook of Human Factors and Ergonomics in Health Care and Patient Safety* (Mahwah, N.J.: Lawrence Erlbaum) p 217-232

[2] Maslach C Leiter M P and Schaufeli W 2008 Measuring Burnout *The Oxford Handbook of Organizational Well Being* (Oxford University Press)

[3] Erwan F Iqbal M Hasanuddin I 2018 The Analysis of Nursing's Work Systems in Relation to Burnout Syndrome (A case study: nurses in RSUDZA, BLUD RSIA, and RSUD Meuraxa, Banda Aceh, Indonesia) *International Conference on Design, Engineering and Computer Sciences*

[4] Jaworek M Marek T Karwowski W Andrzejczak C and Genaidy A 2010 Burnout Syndrome as a Mediator for the Effect of Work-Related Factors on Musculoskeletal Complaints among Hospital Nurses *International Journal of Industrial Ergonomics* **40** 368-375

[5] Goliszek A 2005 *60 Second Manajemen Stress* (Jakarta: PT. Bhuana Ilmu Populer)

[6] Hariyadi U 2006 Burnout Pada Pustakawan [Burnout on Librarian] *Bunga Rampai Perpustakaan dan Informasi Dalam Konteks Budaya* (Jakarta: Universitas Indonesia)

[7] Maslach C 2001 What Have We Learned About Burnout and Health? *Psychology & Health* **16** 5 607-611

[8] Farber B A 1991 *Crisis In Education: Stress and Burnout in the American Teacher* (San Francisco: Jossey-Bass)

[9] Maslach C 2003 *Burnout: The Cost of Caring*

[10] Fatmawati R 2012 Burnout Staf Perpustakaan Bagian Layanan di Badan Perpustakaan dan Arsip Daerah (BPAD) Provinsi DKI Jakarta [Burnout of Service Department Staff of Regional Library and Archives Agency DKI Jakarta, (A master thesis) *Shaut Al-Maktabah: Jurnal Perpustakaan, Arsip dan Dokumentasi* **10** 2 165-176

[11] Tavares K Souza N Da Silva L and Kestenberg C 2014 Ocorrência da Síndrome de Burnout em Enfermeiros Residentes [Prevalence of Burnout Syndrome Among Resident Nurses] *Acta Paulista de Enfermagem* **27** 3 260–265

[12] Maslach C 1982 *Burnout: The Cost of Caring* (Englewood Cliffs, NJ: Prentice-Hall)

[13] Pangastiti N K and Rahardjo M 2011 Analisis Pengaruh Dukungan Sosial Keluarga Terhadap Burnout Pada Perawat Kesehatan Di Rumah Sakit Jiwa (studi pada RSJ Prof. Dr. Soerojo Magelang) [The Influence of Social Support on Family among Health Care Nurses in Psychiatric Hospital] (A doctoral dissertation, Universitas Diponegoro)

AUTHOR INDEX